A GUIDE TO

SQL

Seventh Edition

Philip J. Pratt
Grand Valley State University

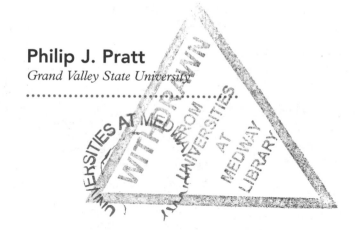

THOMSON
*
COURSE TECHNOLOGY

Australia • Canada • Mexico • Singapore • Spain • United Kingdom • United States

THOMSON

★

COURSE TECHNOLOGY ™

A Guide to SQL, Seventh Edition

by Philip J. Pratt

Senior Vice President, Publisher:
Kristen Duerr

Executive Editor:
Maureen Martin

Product Manager:
Beth Paquin

Developmental Editor:
Jessica Evans

Production Editor:
Brooke Booth

Associate Product Manager:
Mirella Misiaszek

Editorial Assistant:
Jennifer Smith

Marketing Manager:
Karen Seitz

Cover Designer:
Betsy Young

Cover Art:
Rakefet Kennan

Text Designer:
Books by Design

Manufacturing Coordinator:
Laura Burns

COPYRIGHT © 2005 Course Technology, a division of Thomson Learning, Inc.

Thomson Learning™ is a trademark used herein under license.

Printed in Canada

1 2 3 4 5 6 7 8 9 WC 09 08 07 06 05

For more information, contact Course Technology, 25 Thomson Place, Boston, Massachusetts, 02210.

Or find us on the World Wide Web at: www.course.com

Disclaimers

Course Technology reserves the right to revise this publica-tion and make changes from time to time in its content without notice.

Oracle is a registered trade-mark, and Oracle10*g* is a trade-mark or registered trademark of Oracle Corporation and/or its affiliates.

Microsoft is a registered trade-mark of Microsoft Corporation in the United States and/or other countries. Course Technology is an independent entity from the Microsoft Corporation, and is not affiliated with Microsoft in any manner.

MySQL is a registered trade-mark of MySQL AB in the United States, the European Union and other countries.

Some of the product names and company names used in this book have been used for identi-fication purposes only and may be trademarks or registered trademarks of their respective manufacturers and sellers.

ISBN 0-619-21674-3

CONTENTS

Preface   *ix*

1 **Introduction to Premiere Products, Henry Books, and Alexamara Marina Group** 1

Objectives   1
Introduction   2
What is a Database?   2
The Premiere Products Database   2
The Henry Books Database   9
The Alexamara Marina Group Database   16
Summary   21
Key Terms   21
Exercises (Premiere Products)   21
Exercises (Henry Books)   22
Exercises (Alexamara Marina Group)   23

2 **An Introduction to SQL** 25

Objectives   25
Introduction   26
Relational Databases   26
Entities, Attributes, and Relationships   28
Database Creation   31
Running SQL Commands   33
Editing SQL Commands   36
Dropping a Table   41
Data Types   42
Nulls   42
Implementation of Nulls   43
Loading a Table with Data   44
The INSERT Command   44
The INSERT Command with Nulls   48
Viewing Table Data   49
Correcting Errors in the Database   55
Saving SQL Commands   56
Creating the Remaining Database Tables   58
Describing a Table   62
Summary   65
Key Terms   66
Review Questions   66
Exercises (Premiere Products)   68
Exercises (Henry Books)   68
Exercises (Alexamara Marina Group)   70

3 **Single-Table Queries** 73

Objectives   73
Introduction   74
Constructing Simple Queries   75

Retrieving Certain Columns and All Rows   75
Retrieving All Columns and All Rows   76
Using a WHERE Clause   77
Using Compound Conditions   78
Using the BETWEEN Operator   80
Using Computed Columns   81
Using the LIKE Operator   83
Using the IN Operator   84
Sorting   85
Using the ORDER BY Clause   85
Additional Sorting Options   86
Using Functions   87
Using the COUNT Function   87
Using the SUM Function   88
Using the DISTINCT Operator   89
Nesting Queries   91
Subqueries   92
Grouping   93
Using the GROUP BY Clause   94
Using a HAVING Clause   95
HAVING vs. WHERE   95
Nulls   97
Summary   98
Key Terms   99
Review Questions   100
Exercises (Premiere Products)   100
Exercises (Henry Books)   101
Exercises (Alexamara Marina Group)   103

4   Multiple-Table Queries 105
Objectives   105
Introduction   106
Querying Multiple Tables   106
Joining Two Tables   106
Comparing JOIN, IN, and EXISTS   109
Using IN   110
Using EXISTS   110
Using a Subquery within a Subquery   112
A Comprehensive Example   113
Using an Alias   114
Joining a Table to Itself   115
Using a Self-Join on a Primary Key   118
Joining Several Tables   119
Set Operations   121
ALL and ANY   125
Special Operations   128
Inner Join   128
Outer Join   129
Product   131
Summary   132
Key Terms   133
Review Questions   134
Exercises (Premiere Products)   134

Exercises (Henry Books)   136
Exercises (Alexamara Marina Group)   137

5 Updating Data 139
Objectives   139
Introduction   140
Creating a New Table from an Existing Table   140
Changing Existing Data in a Table   141
Adding New Rows to an Existing Table   143
Commit and Rollback   144
Transactions   145
Deleting Existing Rows from a Table   146
Executing a Rollback   147
Changing a Value in a Column to Null   147
Changing a Table's Structure   148
Making Complex Changes   152
Dropping a Table   152
Summary   153
Key Terms   154
Review Questions   154
Exercises (Premiere Products)   155
Exercises (Henry Books)   156
Exercises (Alexamara Marina Group)   157

6 Database Administration 159
Objectives   159
Introduction   160
Views   160
  Row-and-Column Subsets   166
  Joins   167
  Statistics   170
  Dropping a View   170
Security   170
Indexes   173
  Creating an Index   176
  Dropping an Index   177
  Unique Indexes   178
System Catalog   178
Integrity Rules in SQL   181
Summary   184
Key Terms   185
Review Questions   186
Exercises (Premiere Products)   187
Exercises (Henry Books)   189
Exercises (Alexamara Marina Group)   191

7 Reports 193
Objectives   193
Introduction   194
Using Functions   194
  Character Functions   194
  Number Functions   195
  Working with Dates   196

Concatenating Columns   199
Creating and Using Scripts   201
Running the Query for the Report   202
Creating the Data for the Report   204
Changing Column Headings   205
Changing Column Formats in a Report   207
Adding a Title to a Report   209
Grouping Data in a Report   211
Including Totals and Subtotals in a Report   212
Sending the Report to a File   214
Completing the Script to Produce the Report   216
Summary   219
Key Terms   221
Review Questions   221
Exercises (Premiere Products)   222
Exercises (Henry Books)   225
Exercises (Henry Books)   227

8 Embedded SQL 229

Objectives   229
Introduction   230
Using Prompt Variables   231
PL/SQL Programs   232
  Retrieving a Single Row and Column   232
  Using the %TYPE Attribute   234
  Retrieving a Single Row from a Join   235
  Inserting a Row into a Table   236
  Changing a Single Row in a Table   237
  Deleting Rows from a Table   238
  Deleting Rows from Multiple Tables   238
Multiple-Row Select   239
  Using Cursors   240
  Opening a Cursor   240
  Fetching Rows from a Cursor   241
  Closing a Cursor   243
  Complete Program Using a Cursor   243
  More Complex Cursors   244
  Advantages of Cursors   246
  Updating Cursors   247
Error Handling   249
Using SQL in Microsoft Access Programs   249
  Deleting Rows   250
  Running the Code   251
  Updating Rows   252
  Inserting Rows   253
  Finding Rows   253
Summary   255
Key Terms   256
Review Questions   256
Exercises (Premiere Products)   257
Exercises (Henry Books)   258
Exercises (Alexamara Marina Group)   259

A **SQL Reference** 261

    Aliases   262

    ALTER TABLE   262

    Column or Expression List (SELECT Clause)   262

      Computed Columns   263

      The DISTINCT Operator   263

      Functions   263

    COMMIT   263

    Conditions   264

      Simple Conditions   264

      Compound Conditions   264

      BETWEEN Conditions   265

      LIKE Conditions   265

      IN Conditions   265

      EXISTS Conditions   266

      ALL and ANY   266

    CREATE INDEX   266

    CREATE TABLE   267

    CREATE VIEW   267

    Data Types   268

    DELETE Rows   269

    DROP INDEX   269

    DROP TABLE   269

    DROP VIEW   270

    GRANT   270

    INSERT INTO (Query)   271

    INSERT INTO (Values)   271

    Integrity   272

    REVOKE   272

    ROLLBACK   273

    SELECT   273

    Subqueries   274

    UNION, INTERSECT, and MINUS   275

    UPDATE   276

B **"How Do I" Reference** 277

C **Answers to Odd-Numbered Review Questions** 281

    Chapter 1—Introduction to Premiere Products, Henry Books, and Alexamara
        Marina Group   282

    Chapter 2—An Introduction to SQL   282

    Chapter 3—Single-Table Queries   282

    Chapter 4—Multiple-Table Queries   283

    Chapter 5—Updating Data   284

    Chapter 6—Database Administration   284

    Chapter 7—Reports   285

    Chapter 8—Embedded SQL   285

Index   287

PREFACE

Structured Query Language (or SQL, which is pronounced *se-quel*, or *ess-cue-ell*) is a popular computer language that is used by diverse groups such as home computer owners, owners of small businesses, end users in large organizations, and programmers. Although this text uses the SQL implementation in Oracle10g as a vehicle for teaching SQL, its chapter material, examples, and exercises can be completed using any SQL implementation.

A Guide to SQL, Seventh Edition is written for a wide range of teaching levels, from students taking introductory computer science classes to those students in advanced information systems classes. This text can be used for a standalone course on SQL or in conjunction with a database concepts text where students are required to learn SQL.

The chapters in this text should be covered in order. Students should complete the end-of-chapter exercises and the examples within the chapters for maximum learning. Because the content of Chapter 8 assumes that the reader has had instruction or experience in at least one programming language, the instructor should determine whether students will understand its concepts. Students without a programming background will have difficulty understanding the topic of embedded SQL. Instructors can easily omit Chapter 8 from the text in situations where students are not comfortable with programming examples.

Changes in the Seventh Edition

The Seventh Edition builds on the success of previous editions by presenting basic SQL commands in the context of a business that uses SQL to manage orders, parts, customers, and sales reps. The Seventh Edition also contains the changes described in the following sections.

Coverage Includes Oracle10g, Microsoft Access 2003, and MySQL 4.0

Just as in previous editions, this edition uses Oracle as the vehicle to present SQL commands. Like the last edition, this edition addresses SQL in Access by showing the Access versions of the same commands when they differ from the Oracle versions. This new edition also shows MySQL commands when they differ from the Oracle versions. A new pedagogical feature in this edition is the "User" notes, which make it easy for students to identify differences for the SQL implementation they are using.

New Exercises and Case

The Seventh Edition contains new exercises for both the Premiere Products and Henry Books cases. It also includes an additional case, the Alexamara Marina Group.

SQL Reference Appendix

The SQL reference appendix now contains references to the specific pages in the text where the command is discussed to make it easy for students to find additional information when they need to refer back to the section in the book where a topic is covered.

Relationship to *Concepts of Database Management, Fifth Edition*

For database courses featuring SQL, this SQL text can be bundled with *Concepts of Database Management, Fifth Edition* by Pratt and Adamski (Course Technology). The data and pedagogy between the two texts is consistent and the instructor's manuals for both books include feedback and suggestions for using the texts together.

Distinguishing Features

Use of Examples

Each chapter contains multiple examples that use SQL to solve a problem. Following each example, students will read about the commands that are used to solve the stated problem, and then they will see the SQL commands used to arrive at the solution. For most students, learning through examples is the most effective way to master material. For this reason, instructors should encourage students to read the chapters at the computer and input the commands shown in the figures.

Case Studies

A running case study—Premiere Products—is presented in all of the examples within the chapters, and also in the first set of exercises at the end of each chapter. Although the database is small in order to be manageable, the examples and exercises for the Premiere Products database simulate what a real business can accomplish using SQL commands. Using the same case study as examples within the chapter and in the end-of-chapter exercises ensures a high level of continuity to reinforce learning.

A second case study—the Henry Books database—is used in a second set of exercises at the end of each chapter. A third case study—the Alexamara Marina Group database—is used in a third set of exercises at the end of each chapter. The second and third case studies give students a chance to venture out "on their own" without the direct guidance of examples from the text.

Question and Answer Sections

A special type of exercise, called a Q&A, is used throughout the book. These exercises force students to consider special issues and understand important questions before continuing with their study. The answer to each Q&A appears after the question. Students are encouraged to formulate their own answers before reading the ones provided in the text to ensure that they understand new material before proceeding.

"User" Notes for Oracle, Access, and MySQL Users

When an SQL command has a different use or format in Oracle, Access, or MySQL, it appears in a User note. When you encounter a User note for the SQL implementation you are using, you should be sure to read its contents. You might also review the User notes for other SQL implementations so you are aware of the differences that occur from one implementation of SQL to another.

Review Material

A Summary and Key Terms list appear at the end of each chapter, followed by Review Questions that test students' recall of the important points in the chapter and occasionally test their ability to apply what they have learned. The answers to the odd-numbered Review Questions are provided in Appendix C. Each chapter also contains exercises related to the Premiere Products, Henry Books, and Alexamara Marina Group databases.

Appendices

Three appendices appear at the end of this text. Appendix A is an SQL reference that describes the purpose and syntax for the major SQL commands featured in the text. Students can use Appendix A to identify how and when to use important commands quickly. Appendix B includes a "How Do I" reference, which lets students cross-reference the appropriate section in Appendix A by searching for the answer to a question. Appendix C includes answers to the odd-numbered Review Questions.

Instructor Support

The Seventh Edition includes a package of proven supplements for instructors and students. The Instructor's Resources offer a detailed electronic Instructor's Manual, figure files, Microsoft PowerPoint presentations, and the ExamView® Test Bank. The Instructor's Manual includes suggestions and strategies for using this text, as well as answers to Review Questions and solutions to the end-of-chapter exercises. Figure files allow instructors to create their own presentations using figures appearing in the text. Instructors can also take advantage of lecture presentations provided on PowerPoint slides; these presentations follow each chapter's coverage precisely, include chapter figures, and can be customized. ExamView® is a powerful objective-based test generator that enables instructors to create paper, LAN, or Web-based tests from test banks designed specifically for this Course Technology text. Users can utilize the ultra-efficient QuickTest Wizard to create tests in less than five minutes by taking advantage of Course Technology's question banks, or can customize their own exams from scratch.

The Instructor's Resources include copies of the databases for the Premiere Products, Henry Books, and Alexamara Marina Group cases in Microsoft Access 2000/2002/2003 format and script files to create the tables and data in these databases in Oracle and MySQL. These files are provided so instructors have the choice of assigning exercises in which students create the databases used in this text and load them with data, or they can provide the starting Access databases or Oracle or MySQL script files to students to automate and simplify these tasks.

The Instructor's Resources also include text files corresponding to the commands shown in most figures in the text. After opening the text files in Notepad or in a word processor, commands can be copied and pasted in SQL*Plus or SQL*Plus Worksheet rather than being typed directly.

Organization of the Text

The text contains eight chapters and three appendices, which are described in the following sections.

Chapter 1: Introduction to Premiere Products, Henry Books, and Alexamara Marina Group

Chapter 1 introduces the three database cases that are used throughout the text: Premiere Products, Henry Books, and Alexamara Marina Group. Many Q&A exercises are provided throughout the chapter to ensure that students understand how to manipulate the database on paper before they begin working in SQL.

Chapter 2: An Introduction to SQL

In Chapter 2, students learn about important concepts and terminology associated with relational databases. Then they create and run SQL commands to create tables, use data types, and add rows to tables. Chapter 2 also discusses the role and use of nulls.

Chapter 3: Single-Table Queries

Chapter 3 is the first of two chapters on using SQL commands to query a database. The queries in Chapter 3 all involve single tables. Included in this chapter are discussions of simple and compound conditions; computed columns; the SQL BETWEEN, LIKE, and IN operators; using SQL functions; nesting queries; grouping data; and retrieving columns with null values.

Chapter 4: Multiple-Table Queries

Chapter 4 completes the discussion of querying a database by demonstrating queries that join more than one table. Included in this chapter are discussions of the SQL IN and EXISTS operators, nested subqueries, using aliases, joining a table to itself, SQL set operations, and the use of the ALL and ANY operators. The chapter also includes coverage of various types of joins.

Chapter 5: Updating Data

In Chapter 5, students learn how to use the SQL COMMIT, ROLLBACK, UPDATE, INSERT, and DELETE commands to update table data. Students also learn how to create a new table from an existing table and how to change the structure of a table. The chapter also includes coverage of transactions, including both their purpose and implementation.

Chapter 6: Database Administration

Chapter 6 covers the database administration features of SQL, including the use of views; granting and revoking database privileges to users; creating, dropping, and using an index; using and obtaining information from the system catalog; and using integrity constraints to control data entry. In addition, for those Access tasks that you would not typically use SQL to complete, the Seventh Edition also includes the approach most commonly used in Access.

Chapter 7: Reports

Chapter 7 begins with a discussion of some important SQL functions that act on single rows. Chapter 7 then teaches students how to create basic and complex reports based on data in a table or view. Students will learn how to concatenate data, create a view for a report, change report column headings and formats, and add report titles. Students also will include totals and subtotals in a report and group data. The topics of scripts and spooling also are discussed.

Chapter 8: Embedded SQL

Chapter 8 uses PL/SQL to cover the process of embedding SQL commands in another language. Included in this chapter are discussions of the use of embedded SQL to insert new rows and change and delete existing rows. Also included is a discussion of how to retrieve single rows using embedded SQL commands and how to use cursors to retrieve multiple rows. The chapter concludes with a section showing some techniques for using SQL in Access VBA.

Appendix A: SQL Reference

Appendix A includes a command reference for all the major SQL clauses and operators that are featured in the chapters. Students can use Appendix A as a quick resource when constructing commands. Each command includes a short description, a table that shows the required and optional clauses and operators, and an example and its results. It also contains a reference to the pages in the text where the command is covered.

Appendix B: "How Do I" Reference

Appendix B provides students with an opportunity to ask a question, such as "How do I delete rows?," and to identify the appropriate section in Appendix A to use to find the answer. Appendix B is extremely valuable when students know what task they want to accomplish, but can't remember the exact SQL command they need.

Appendix C: Answers to Odd-Numbered Review Questions

Answers to the odd-numbered Review Questions in each chapter appear in this appendix so students can make sure that they are completing the Review Questions correctly.

General Notes to the Student

You can download the databases used in this text from *www.course.com*. The Access data files for this book include three Access databases (Premiere Products.mdb, Henry Books.mdb, and Alexamara Marina Group.mdb), which you can open in Access 2000, 2002, or 2003.

The Data Disk also includes script files for Oracle and MySQL that you can use to create or drop the Alexamara Marina Group, Henry Books, and Premiere Products databases. The script files saved in the Oracle folder on the Data Disk have the following functions:

- **Oracle-Alexamara.sql:** Creates all the tables in the Alexamara Marina Group database and adds all the data. Run this script file to create the Alexamara Marina Group database. (*Note:* This script file assumes you have not previously created any of the tables in the database. If you have created any of the tables, you should run the Oracle-DropAlexamara.sql script prior to running the Oracle-Alexamara.sql script.)

- **Oracle-Henry.sql:** Creates all the tables in the Henry Books database and adds all the data. Run this script file to create the Henry Books database. (*Note:* This script file assumes you have not previously created any of the tables in the database. If you have created any of the tables, you should run the Oracle-DropHenry.sql script prior to running the Oracle-Henry.sql script.)

- **Oracle-Premiere.sql:** Creates all the tables in the Premiere Products database and adds all the data. Run this script file to create the Premiere Products database. (*Note:* This script file assumes you have not previously created any of the tables in the database. If you have created any of the tables, you should run the Oracle-DropPremiere.sql script prior to running the Oracle-Premiere.sql script.)

- **Oracle-DropAlexamara.sql:** Drops (deletes) all the tables and data in the Alexamara Marina Group database.

- **Oracle-DropHenry.sql:** Drops (deletes) all the tables and data in the Henry Books database.

- **Oracle-DropPremiere.sql:** Drops (deletes) all the tables and data in the Premiere Products database.

The script files saved in the MySQL folder on the Data Disk have the following functions:

- **MySQL-Alexamara:** Creates all the tables in the Alexamara Marina Group database and adds all the data. Run this script file to create the Alexamara Marina Group database. (*Note:* This script file assumes you have not previously created any of the tables in the database. If you have created any of the tables, you should run the MySQL-DropAlexamara script prior to running the MySQL-Alexamara script.)

- **MySQL-Henry:** Creates all the tables in the Henry Books database and adds all the data. Run this script file to create the Henry Books database. (*Note:* This script file assumes you have not previously created any of the tables in the database. If you have created any of the tables, you should run the MySQL-DropHenry script prior to running the MySQL-Henry script.)

- **MySQL-Premiere:** Creates all the tables in the Premiere Products database and adds all the data. Run this script file to create the Premiere Products database. (*Note:* This script file assumes you have not previously created any of the tables in the database. If you have created any of the tables, you should run the MySQL-DropPremiere script prior to running the MySQL-Premiere script.)

- **MySQL-DropAlexamara:** Drops (deletes) all the tables and data in the Alexamara Marina Group database.

- **MySQL-DropHenry:** Drops (deletes) all the tables and data in the Henry Books database.

- **MySQL-DropPremiere:** Drops (deletes) all the tables and data in the Premiere Products database.

For details on running script files in Oracle or MySQL, check with your instructor. You can also refer to Chapters 2 and 7 in the text for information about creating and using scripts.

The Data Disk also contains script files that create the final report in Chapter 7 and several PL/SQL programs in Chapter 8. You can run the script files instead of typing the programs directly. The script files correspond to the following figures in the text and are stored in the Chapter07 and Chapter08 folders on the Data Disk:

- Figure 7.23 (F7-23.sql)

- Figure 8.2 (F8-02.sql)

- Figure 8.4 (F8-04.sql)

- Figure 8.5 (F8-05.sql)

- Figure 8.7 (F8-07.sql)

- Figure 8.9 (F8-09.sql)

- Figure 8.10 (F8-10.sql)

- Figure 8.11 (F8-11.sql)

- Figure 8.19 (F8-19.sql)

- Figure 8.21 (F8-21.sql)

- Figure 8.23 (F8-23.sql)

- Figure 8.24 (F8-24.sql)

Embedded Questions

In many places, you'll find Q&A sections to ensure that you understand some crucial material before you proceed. In some cases, the questions are designed to give you the chance to consider some special concept in advance of its actual presentation. In all cases, the answer to each question appears immediately after the question. You can simply read the question and its answer, but you will benefit from taking time to determine the answer to the question before checking your answer against the one given in the text.

End-of-Chapter Material

The end-of-chapter material consists of a Summary, a Key Terms list, Review Questions, and exercises for the Premiere Products, Henry Books, and Alexamara Marina Group databases. The Summary briefly describes the material covered in the chapter. The Review Questions require you to recall and apply the important material in the chapter. The answers to the odd-numbered Review Questions appear in Appendix C so you can check your progress. The Premiere Products, Henry Books, and Alexamara Marina Group exercises test your knowledge of the chapter material; your instructor will assign one or more of these exercises for you to complete.

Acknowledgments

I would like to acknowledge several individuals for their contributions in the preparation of this text. I appreciate the efforts of the following individuals who reviewed the manuscript and made many helpful suggestions: Vickee Stedham, St. Petersburg College; Bill Kloepfer, Golden Gate University; Georgia Brown, Northern Illinois University; Gary Savard, Champlain College; Stephen Cerovski, Coleman College; Ricardo Herrera, Vanier College and Concordia University; Eugenia Fernandez, Indiana University-Purdue University Indianapolis; Danny Yakimchuk, University College of Cape Breton; Paul Leidig, Grand Valley State University; Misty Vermaat, Purdue University Calumet; Lorna Bowen St. George, Old Dominion University; and George Federman, Santa Barbara Community College.

The efforts of the following members of the staff at Course Technology, Inc. have been invaluable and have made this text possible: Maureen Martin, Acquisitions Editor; Beth Paquin, Product Manager; Brooke Booth, Production Editor; and Quality Assurance testers Chris Scriver, Marianne Snow, and Serge Palladino.

I have had the great pleasure to work with an absolutely amazing Developmental Editor, Jessica Evans, on several books. Thanks for all your efforts, Jess. You're the best!

CHAPTER 1

Introduction to Premiere Products, Henry Books, and Alexamara Marina Group

OBJECTIVES

- Introduce Premiere Products, a company whose database is used as the basis for many of the examples throughout the text

- Introduce Henry Books, a company whose database is used as a case that runs throughout the text

- Introduce Alexamara Marina Group, a company whose database is used as an additional case that runs throughout the text

Introduction

In this chapter, you will examine the database requirements of Premiere Products, a company that will be used in the examples throughout the text. Then you will examine the database requirements for Henry Books and Alexamara Marina Group, whose databases are featured in the exercises that appear at the end of each chapter.

■ What is a Database?

Throughout this text, you will work with databases for three organizations: Premiere Products, Henry Books, and Alexamara Marina Group. A **database** is a structure that contains different categories of information and the relationships between these categories. The Premiere Products database, for example, contains information about categories such as sales reps, customers, orders, and parts. The Henry Books database contains information about categories such as books, publishers, authors, and branches. The Alexamara Marina Group database contains information about categories such as marinas, slips and the boats in them, service categories, and service requests.

Each database also contains relationships between categories. For example, the Premiere Products database contains information that relates sales reps to the customers they represent and customers to the orders they have placed. The Henry Books database contains information that relates publishers to the books they publish and authors to the books they have written. The Alexamara Marina Group database contains information that relates the boats in the slips at the marina to the owners of the boats.

As you work through the chapters in this text, you will learn more about these databases and how to view the information they contain. As you read each chapter, you will see examples from the Premiere Products database. At the end of each chapter, your instructor might assign the exercises for the Premiere Products, Henry Books, or Alexamara Marina Group databases.

■ The Premiere Products Database

The management of Premiere Products, a distributor of appliances, housewares, and sporting goods, has determined that the company's recent growth makes it no longer feasible to maintain customer, order, and inventory data using its manual systems. With the data stored in a database management system, management will be able to ensure that the data is current and more accurate than in the present manual systems. In addition, managers will be able to obtain answers to their questions concerning the data in the database easily and quickly, with the option of producing a variety of useful reports.

Management has determined that Premiere Products must maintain the following information about its sales representatives (sales reps), customers, and parts inventory in the new database:

- The number, last name, first name, address, total commission, and commission rate for each sales rep.

- The customer number, name, address, current balance, and credit limit for each customer, as well as the number of the sales rep who represents the customer.

- The part number, description, number of units on hand, item class, number of the warehouse where the item is stored, and unit price for each part in inventory.

Premiere Products must also store information about orders. Figure 1.1 shows a sample order.

FIGURE 1.1   Sample order

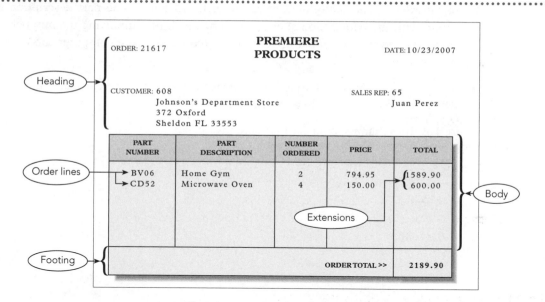

The sample order shown in Figure 1.1 has three sections:

- The heading (top) of the order contains the order number and date; the customer's number, name, and address; and the sales rep's number and name.

- The body of the order contains one or more order lines, sometimes called line items. Each order line contains a part number, a part description, the number of units of the part ordered, and the quoted price for the part. Each order line also contains a total, usually called an extension, which is the result of multiplying the number ordered by the quoted price.

- Finally, the footing (bottom) of the order contains the order total.

Premiere Products must also store the following items for each customer's order:

- For each order, it must store the order number, the date the order was placed, and the number of the customer that placed the order. The customer's name and address and the number of the sales rep who represents the customer are stored with customer information. The name of the sales rep is stored with the sales rep information.

- For each order, it must store the order number, the part number, the number of units ordered, and the quoted price for each order line. The part description is stored with the information about parts. The result of multiplying the number of units ordered by the quoted price is not stored because the computer can calculate it when needed.

- The overall order total is not stored. Instead, the computer calculates the total whenever an order is printed or displayed on the screen.

Figure 1.2 shows sample data for Premiere Products.

FIGURE 1.2   Sample data for Premiere Products

REP

REP_NUM	LAST_NAME	FIRST_NAME	STREET	CITY	STATE	ZIP	COMMISSION	RATE
20	Kaiser	Valerie	624 Randall	Grove	FL	33321	$20,542.50	0.05
35	Hull	Richard	532 Jackson	Sheldon	FL	33553	$39,216.00	0.07
65	Perez	Juan	1626 Taylor	Fillmore	FL	33336	$23,487.00	0.05

CUSTOMER

CUSTOMER_NUM	CUSTOMER_NAME	STREET	CITY	STATE	ZIP	BALANCE	CREDIT_LIMIT	REP_NUM
148	Al's Appliance and Sport	2837 Greenway	Fillmore	FL	33336	$6,550.00	$7,500.00	20
282	Brookings Direct	3827 Devon	Grove	FL	33321	$431.50	$10,000.00	35
356	Ferguson's	382 Wildwood	Northfield	FL	33146	$5,785.00	$7,500.00	65
408	The Everything Shop	1828 Raven	Crystal	FL	33503	$5,285.25	$5,000.00	35
462	Bargains Galore	3829 Central	Grove	FL	33321	$3,412.00	$10,000.00	65
524	Kline's	838 Ridgeland	Fillmore	FL	33336	$12,762.00	$15,000.00	20
608	Johnson's Department Store	372 Oxford	Sheldon	FL	33553	$2,106.00	$10,000.00	65
687	Lee's Sport and Appliance	282 Evergreen	Altonville	FL	32543	$2,851.00	$5,000.00	35
725	Deerfield's Four Seasons	282 Columbia	Sheldon	FL	33553	$248.00	$7,500.00	35
842	All Season	28 Lakeview	Grove	FL	33321	$8,221.00	$7,500.00	20

ORDERS

ORDER_NUM	ORDER_DATE	CUSTOMER_NUM
21608	10/20/2007	148
21610	10/20/2007	356
21613	10/21/2007	408
21614	10/21/2007	282
21617	10/23/2007	608
21619	10/23/2007	148
21623	10/23/2007	608

ORDER_LINE

ORDER_NUM	PART_NUM	NUM_ORDERED	QUOTED_PRICE
21608	AT94	11	$21.95
21610	DR93	1	$495.00
21610	DW11	1	$399.99
21613	KL62	4	$329.95
21614	KT03	2	$595.00
21617	BV06	2	$794.95
21617	CD52	4	$150.00
21619	DR93	1	$495.00
21623	KV29	2	$1,290.00

(continued)

FIGURE 1.2   Sample data for Premiere Products (continued)

PART

PART_NUM	DESCRIPTION	ON_HAND	CLASS	WAREHOUSE	PRICE
AT94	Iron	50	HW	3	$24.95
BV06	Home Gym	45	SG	2	$794.95
CD52	Microwave Oven	32	AP	1	$165.00
DL71	Cordless Drill	21	HW	3	$129.95
DR93	Gas Range	8	AP	2	$495.00
DW11	Washer	12	AP	3	$399.99
FD21	Stand Mixer	22	HW	3	$159.95
KL62	Dryer	12	AP	1	$349.95
KT03	Dishwasher	8	AP	3	$595.00
KV29	Treadmill	9	SG	2	$1,390.00

In the REP table, you see that there are three sales reps, whose numbers are 20, 35, and 65. The name of sales rep 20 is Valerie Kaiser. Her street address is 624 Randall. She lives in Grove, FL, and her zip code is 33321. Her total commission is $20,542.50, and her commission rate is 5% (0.05).

Premiere Products has 10 customers that are identified with the numbers 148, 282, 356, 408, 462, 524, 608, 687, 725, and 842. The name of customer number 148 is Al's Appliance and Sport. This customer's address is 2837 Greenway in Fillmore, FL, with a zip code of 33336. The customer's current balance is $6,550.00, and its credit limit is $7,500.00. The number 20 in the REP_NUM column indicates that Al's Appliance and Sport is represented by sales rep 20 (Valerie Kaiser).

Skipping to the table named PART, you see that there are 10 parts, whose part numbers are AT94, BV06, CD52, DL71, DR93, DW11, FD21, KL62, KT03, and KV29. Part AT94 is an iron, and the company has 50 units of this part on hand. Irons are in item class HW (housewares) and are stored in warehouse 3. The price of an iron is $24.95. Other item classes are AP (appliances) and SG (sporting goods).

Moving back to the table named ORDERS, you see that there are seven orders, which are identified with the numbers 21608, 21610, 21613, 21614, 21617, 21619, and 21623. Order number 21608 was placed on October 20, 2007 by customer 148 (Al's Appliance and Sport).

Note: In some database systems, the word "order" has a special purpose. Having a table named ORDER could cause problems in such systems. For this reason, Premiere Products uses the table name ORDERS instead of ORDER.

The table named ORDER_LINE might seem strange at first glance. Why do you need a separate table for the order lines? Could they be included in the ORDERS table? The answer is technically yes. The table named ORDERS could be structured as shown in Figure 1.3. Notice that this table contains the same orders as shown in Figure 1.2, with the same dates and customer numbers. In addition, each table row in Figure 1.3 contains all the order lines for a given order. Examining the fifth row, for example, you see that order 21617 has two order lines. One of these order lines is for two BV06 parts at $794.95 each, and the other order line is for four CD52 parts at $150.00 each.

FIGURE 1.3 Alternative ORDERS table structure

ORDERS

ORDER_NUM	ORDER_DATE	CUSTOMER_NUM	PART_NUM	NUM_ORDERED	QUOTED_PRICE
21608	10/20/2007	148	AT94	11	$21.95
21610	10/20/2007	356	DR93	1	$495.00
			DW11	1	$399.99
21613	10/21/2007	408	KL62	4	$329.95
21614	10/21/2007	282	KT03	2	$595.00
21617	10/23/2007	608	BV06	2	$794.95
			CD52	4	$150.00
21619	10/23/2007	148	DR93	1	$495.00
21623	10/23/2007	608	KV29	2	$1,290.00

QUESTION How is the information from Figure 1.2 represented in Figure 1.3?

ANSWER Examine the ORDER_LINE table shown in Figure 1.2 and note the sixth and seventh rows. The sixth row indicates that there is an order line on order 21617 for two BV06 parts at $794.95 each. The seventh row indicates that there is an order line on order 21617 for four CD52 parts at $150.00 each. Thus, the information that you find in Figure 1.3 is represented in Figure 1.2 in two separate rows rather than in one row.

It might seem inefficient to use two rows to store information that could be represented in one row. There is a problem, however, with the arrangement shown in Figure 1.3—the table is more complicated. In Figure 1.2, there is a single entry at each location in the table. In Figure 1.3, some of the individual positions within the table contain multiple entries, making it difficult to track the information between columns. In the row for order number 21617, for example, it is crucial to know that the BV06 corresponds to the 2 in the

NUM_ORDERED column (not the 4) and that it corresponds to the $794.95 in the QUOTED_PRICE column (not the $150.00). In addition, a more complex table raises practical issues, such as:

- How much room do you allow for these multiple entries?

- What if an order has more order lines than you have allowed room for?

- For a given part, how do you determine which orders contain order lines for that part?

Although none of these problems is unsolvable, they do add a level of complexity that is not present in the arrangement shown in Figure 1.2. In Figure 1.2, there are no multiple entries to worry about, it doesn't matter how many order lines exist for any order, and finding every order that contains an order line for a given part is easy (just look for all order lines with the given part number in the PART_NUM column). In general, this simpler structure is preferable, and that is why order lines appear in a separate table.

To test your understanding of the Premiere Products data, use Figure 1.2 to answer the following questions.

QUESTION   What are the numbers of the customers represented by Valerie Kaiser?

ANSWER   148, 524, and 842. (Look up the REP_NUM value of Valerie Kaiser in the REP table and obtain the number 20. Then find all customers in the CUSTOMER table that have the number 20 in the REP_NUM column.)

QUESTION   What is the name of the customer that placed order 21610, and what is the name of the rep who represents this customer?

ANSWER   Ferguson's is the customer, Juan Perez is the sales rep. (Look up the CUSTOMER_NUM value in the ORDERS table for order number 21610 and obtain the number 356. Then find the customer in the CUSTOMER table with the CUSTOMER_NUM value of 356. Using the REP_NUM value, which is 65, find the name of the rep in the REP table.)

QUESTION   List all parts that appear in order 21610. For each part, give the description, number ordered, and quoted price.

ANSWER   Part number: DR93; part description: Gas Range; number ordered: 1; and quoted price: $495.00. Also, part number: DW11; part description: Washer; number ordered: 1; and quoted price: $399.99. (Look up each ORDER_LINE table row in which the order number is 21610. Each of these rows contains a part number, the number ordered, and the quoted price. Use the part number to look up the corresponding part description in the PART table.)

QUESTION   Why is the QUOTED_PRICE column part of the ORDER_LINE table? Can't you just use the part number and look up the price in the PART table?

ANSWER   If the QUOTED_PRICE column didn't appear in the ORDER_LINE table, you would need to obtain the price for a part on an order line by looking up the price in the PART table. Although this might not be a bad practice, it prevents Premiere Products from charging different prices to different customers for the same part. Because Premiere Products wants the flexibility to quote and charge different prices to different customers, the QUOTED_PRICE column is included in the ORDER_LINE table. If you examine the ORDER_LINE table, you will see cases in which the quoted price matches the actual price in the PART table and cases in which it differs. For example, in order number 21608, Al's Appliance and Sport bought 11 irons, and Premiere Products charged only $21.95 per iron, rather than the regular price of $24.95.

■ The Henry Books Database

Ray Henry is the owner of a bookstore chain named Henry Books. Like the management of Premiere Products, Ray has decided to store his data in a database. He wants to achieve the same benefits; that is, he wants to ensure that his data is current and accurate. He also needs to create forms to interact with the data and to produce reports from that data. In addition, he wants to be able to ask questions concerning the data and to obtain answers to these questions easily and quickly.

In running his chain of bookstores, Ray gathers and organizes information about branches, publishers, authors, and books. Figure 1.4 shows sample branch and publisher data for Henry Books. Each branch has a number that uniquely identifies the branch. In addition, Ray tracks the branch's name, location, and number of employees. Each publisher has a code that uniquely identifies the publisher. In addition, Ray tracks the publisher's name and city.

FIGURE 1.4   Sample branch and publisher data for Henry Books

BRANCH

BRANCH_NUM	BRANCH_NAME	BRANCH_LOCATION	NUM_EMPLOYEES
1	Henry Downtown	16 Riverview	10
2	Henry On The Hill	1289 Bedford	6
3	Henry Brentwood	Brentwood Mall	15
4	Henry Eastshore	Eastshore Mall	9

PUBLISHER

PUBLISHER_CODE	PUBLISHER_NAME	CITY
AH	Arkham House	Sauk City WI
AP	Arcade Publishing	New York
BA	Basic Books	Boulder CO
BP	Berkley Publishing	Boston
BY	Back Bay Books	New York
CT	Course Technology	Boston
FA	Fawcett Books	New York
FS	Farrar Straus and Giroux	New York
HC	HarperCollins Publishers	New York
JP	Jove Publications	New York
JT	Jeremy P. Tarcher	Los Angeles
LB	Lb Books	New York
MP	McPherson and Co.	Kingston
PE	Penguin USA	New York
PL	Plume	New York
PU	Putnam Publishing Group	New York
RH	Random House	New York
SB	Schoken Books	New York
SC	Scribner	New York
SS	Simon and Schuster	New York
ST	Scholastic Trade	New York
TA	Taunton Press	Newtown CT
TB	Tor Books	New York
TH	Thames and Hudson	New York
TO	Touchstone Books	Westport CT
VB	Vintage Books	New York
WN	W.W. Norton	New York
WP	Westview Press	Boulder CO

Figure 1.5 shows sample author data for Henry Books. Each author has a number that uniquely identifies the author. In addition, Ray records each author's last and first names.

FIGURE 1.5   Sample author data for Henry Books

AUTHOR

AUTHOR_NUM	AUTHOR_LAST	AUTHOR_FIRST
1	Morrison	Toni
2	Solotaroff	Paul
3	Vintage	Vernor
4	Francis	Dick
5	Straub	Peter
6	King	Stephen
7	Pratt	Philip
8	Chase	Truddi
9	Collins	Bradley
10	Heller	Joseph
11	Wills	Gary
12	Hofstadter	Douglas R.
13	Lee	Harper
14	Ambrose	Stephen E.
15	Rowling	J.K.
16	Salinger	J.D.
17	Heaney	Seamus
18	Camus	Albert
19	Collins, Jr.	Bradley
20	Steinbeck	John
21	Castelman	Riva
22	Owen	Barbara
23	O'Rourke	Randy
24	Kidder	Tracy
25	Schleining	Lon

Figure 1.6 shows sample book data for Henry Books. Each book has a code that uniquely identifies the book. For each book, Ray also tracks the title, publisher, type of book, price, and whether the book is a paperback.

FIGURE 1.6 Sample book data for Henry Books

BOOK

BOOK_CODE	TITLE	PUBLISHER_CODE	TYPE	PRICE	PAPERBACK
0180	A Deepness in the Sky	TB	SFI	$7.19	Yes
0189	Magic Terror	FA	HOR	$7.99	Yes
0200	The Stranger	VB	FIC	$8.00	Yes
0378	Venice	SS	ART	$24.50	No
079X	Second Wind	PU	MYS	$24.95	No
0808	The Edge	JP	MYS	$6.99	Yes
1351	Dreamcatcher: A Novel	SC	HOR	$19.60	No
1382	Treasure Chests	TA	ART	$24.46	No
138X	Beloved	PL	FIC	$12.95	Yes
2226	Harry Potter and the Prisoner of Azkaban	ST	SFI	$13.96	No
2281	Van Gogh and Gauguin	WP	ART	$21.00	No
2766	Of Mice and Men	PE	FIC	$6.95	Yes
2908	Electric Light	FS	POE	$14.00	No
3350	Group: Six People in Search of a Life	BP	PSY	$10.40	Yes
3743	Nine Stories	LB	FIC	$5.99	Yes
3906	The Soul of a New Machine	BY	SCI	$11.16	Yes
5163	Travels with Charley	PE	TRA	$7.95	Yes
5790	Catch-22	SC	FIC	$12.00	Yes
6128	Jazz	PL	FIC	$12.95	Yes
6328	Band of Brothers	TO	HIS	$9.60	Yes
669X	A Guide to SQL	CT	CMP	$37.95	Yes
6908	Franny and Zooey	LB	FIC	$5.99	Yes
7405	East of Eden	PE	FIC	$12.95	Yes
7443	Harry Potter and the Goblet of Fire	ST	SFI	$18.16	No
7559	The Fall	VB	FIC	$8.00	Yes
8092	Godel, Escher, Bach	BA	PHI	$14.00	Yes
8720	When Rabbit Howls	JP	PSY	$6.29	Yes
9611	Black House	RH	HOR	$18.81	No
9627	Song of Solomon	PL	FIC	$14.00	Yes
9701	The Grapes of Wrath	PE	FIC	$13.00	Yes
9882	Slay Ride	JP	MYS	$6.99	Yes
9883	The Catcher in the Rye	LB	FIC	$5.99	Yes
9931	To Kill a Mockingbird	HC	FIC	$18.00	No

To check your understanding of the relationship between publishers and books, answer the following questions.

QUESTION   Who published *Jazz*? Which books did Jove Publications publish?

ANSWER   Plume published *Jazz*. In the row in the BOOK table for *Jazz* (see Figure 1.6), find the publisher code PL. Examining the PUBLISHER table (see Figure 1.4), you see that PL is the code assigned to Plume.

Jove Publications published *The Edge*, *When Rabbit Howls*, and *Slay Ride*. To find the books published by Jove Publications, find its code (JP) in the PUBLISHER table. Next, find all records in the BOOK table for which the publisher code is JP.

The table named WROTE, as shown in Figure 1.7, is used to relate books and authors. The SEQUENCE field indicates the order in which the authors of a particular book are listed on the cover. The table named INVENTORY in the same figure is used to indicate the number of copies of a particular book that are currently on hand at a particular branch of Henry Books. The first row, for example, indicates that there are two copies of the book with the code 0180 at branch 1.

FIGURE 1.7   Sample data that relates books to authors and books to branches for Henry Books

WROTE

BOOK_CODE	AUTHOR_NUM	SEQUENCE
0180	3	1
0189	5	1
0200	18	1
0378	11	1
079X	4	1
0808	4	1
1351	6	1
1382	23	2
1382	25	1
138X	1	1
2226	15	1
2281	9	2
2281	19	1
2766	20	1

INVENTORY

BOOK_CODE	BRANCH_NUM	ON_HAND
0180	1	2
0189	2	2
0200	1	1
0200	2	3
0378	3	2
079X	2	1
079X	3	2
079X	4	3
0808	2	1
1351	2	4
1351	3	2
1382	2	1
138X	2	3
2226	1	3

(continued)   (continued)

WROTE

BOOK_CODE	AUTHOR_NUM	SEQUENCE
2908	17	1
3350	2	1
3743	16	1
3906	24	1
5163	20	1
5790	10	1
6128	1	1
6328	14	1
669X	7	1
6908	16	1
7405	20	1
7443	15	1
7559	18	1
8092	12	1
8720	8	1
9611	5	2
9611	6	1
9627	1	1
9701	20	1
9882	4	1
9883	16	1
9931	13	1

INVENTORY

BOOK_CODE	BRANCH_NUM	ON_HAND
2226	3	2
2226	4	1
2281	4	3
2766	3	2
2908	1	3
2908	4	1
3350	1	2
3743	2	1
3906	2	1
3906	3	2
5163	1	1
5790	4	2
6128	2	4
6128	3	3
6328	2	2
669X	1	1
6908	2	2
7405	3	2
7443	4	1
7559	2	2
8092	3	1
8720	1	3
9611	1	2
9627	3	5
9627	4	2
9701	1	2
9701	2	1
9701	3	3
9701	4	2
9882	3	3
9883	2	3
9883	4	2
9931	1	2

To check your understanding of the relationship between authors and books, answer the following questions.

QUESTION Who wrote *Black House*? (Make sure to list the authors in the correct order.) Which books did Toni Morrison write?

ANSWER Stephen King and Peter Straub wrote *Black House*. First examine the BOOK table (see Figure 1.6) to find the book code for *Black House* (9611). Next, look for all rows in the WROTE table in which the book code is 9611. There are two such rows. In one row, the author number is 5, and in the other, it is 6. Then, look in the AUTHOR table to find the authors who have been assigned the numbers 5 and 6. The answers are Peter Straub (5) and Stephen King (6). The sequence number for author number 5 is 2 and the sequence number for author number 6 is 1. Thus, listing the authors in the proper order results in Stephen King and Peter Straub.

Toni Morrison wrote *Beloved*, *Jazz*, and *Song of Solomon*. To find the books written by Toni Morrison, look up her author number in the AUTHOR table (it is 1). Then look for all rows in the WROTE table for which the author number is 1. There are three such rows. The corresponding book codes are 138X, 6128, and 9627. Looking up these codes in the BOOK table, you find that Toni Morrison wrote *Beloved*, *Jazz*, and *Song of Solomon*.

QUESTION A customer in branch 1 wishes to purchase *The Soul of a New Machine*. Is it currently in stock at branch 1?

ANSWER No. Looking up the code for *The Soul of a New Machine* in the BOOK table, you find it is 3906. To find out how many copies are in stock at branch 1, look for a row in the INVENTORY table with 3906 in the BOOK_CODE column and 1 in the BRANCH_NUM column. Because there is no such row, branch 1 doesn't have any copies of *The Soul of a New Machine*.

QUESTION You would like to obtain a copy of *The Soul of a New Machine* for this customer. Which other branches currently have it in stock, and how many copies does each branch have?

ANSWER Branch 2 has one copy, and branch 3 has two copies. You already know that the code for *The Soul of a New Machine* is 3906. (If you didn't know the book code, you would look it up in the BOOK table.) To find out which branches currently have copies, look for rows in the INVENTORY table with 3906 in the BOOK_CODE column. There are two such rows. The first row indicates that branch number 2 currently has one copy. The second row indicates that branch number 3 currently has two copies.

■ The Alexamara Marina Group Database

Alexamara Marina Group offers in-water storage to boat owners by providing boat slips that boat owners can rent on an annual basis. Alexamara owns two marinas: Alexamara East and Alexamara Central. Each marina has several boat slips available. Alexamara also provides a variety of boat repair and maintenance services to the boat owners who rent the slips. Alexamara stores the data it needs to manage its operations in a relational database containing the tables described in the following section.

Alexamara stores information about its two marinas in the MARINA table shown in Figure 1.8. A marina number uniquely identifies each marina. The table also includes the marina name, street address, city, state, and zip code.

FIGURE 1.8 MARINA table

MARINA

MARINA_NUM	NAME	ADDRESS	CITY	STATE	ZIP
1	Alexamara East	108 2nd Ave.	Brinman	FL	32273
2	Alexamara Central	283 Branston	W. Brinman	FL	32274

Alexamara stores information about the boat owners to whom it rents slips in the OWNER table shown in Figure 1.9. An owner number that consists of two uppercase letters followed by a two-digit number uniquely identifies each owner. For each owner, the table also includes the last name, first name, address, city, state, and zip code.

FIGURE 1.9 OWNER table

OWNER

OWNER_NUM	LAST_NAME	FIRST_NAME	ADDRESS	CITY	STATE	ZIP
AD57	Adney	Bruce and Jean	208 Citrus	Bowton	FL	31313
AN75	Anderson	Bill	18 Wilcox	Glander Bay	FL	31044
BL72	Blake	Mary	2672 Commodore	Bowton	FL	31313
EL25	Elend	Sandy and Bill	462 Riverside	Rivard	FL	31062
FE82	Feenstra	Daniel	7822 Coventry	Kaleva	FL	32521
JU92	Juarez	Maria	8922 Oak	Rivard	FL	31062
KE22	Kelly	Alyssa	5271 Waters	Bowton	FL	31313
NO27	Norton	Peter	2811 Lakewood	Lewiston	FL	32765
SM72	Smeltz	Becky and Dave	922 Garland	Glander Bay	FL	31044
TR72	Trent	Ashton	922 Crest	Bay Shores	FL	30992

Each of the marinas contains slips that are identified by slip numbers. Marina 1 (Alexamara East) has two sections named A and B. Slips are numbered within each section. Thus, slip numbers at marina 1 consist of the letter A or B followed by a number (for example, A3 or B2). At marina 2 (Alexamara Central), a number (1, 2, 3) identifies each slip.

Information about the slips in the marinas is contained in the MARINA_SLIP table shown in Figure 1.10. Each row in the table contains a slip ID that identifies the particular slip. The table also contains the marina number and slip number, the length of the slip (in feet), the annual rental fee, the name of the boat currently occupying the slip, the type of boat, and the boat owner's number.

FIGURE 1.10 MARINA_SLIP table

MARINA_SLIP

SLIP_ID	MARINA_NUM	SLIP_NUM	LENGTH	RENTAL_FEE	BOAT_NAME	BOAT_TYPE	OWNER_NUM
1	1	A1	40	$3,800.00	Anderson II	Sprite 4000	AN75
2	1	A2	40	$3,800.00	Our Toy	Ray 4025	EL25
3	1	A3	40	$3,600.00	Escape	Sprite 4000	KE22
4	1	B1	30	$2,400.00	Gypsy	Dolphin 28	JU92
5	1	B2	30	$2,600.00	Anderson III	Sprite 3000	AN75
6	2	1	25	$1,800.00	Bravo	Dolphin 25	AD57
7	2	2	25	$1,800.00	Chinook	Dolphin 22	FE82
8	2	3	25	$2,000.00	Listy	Dolphin 25	SM72
9	2	4	30	$2,500.00	Mermaid	Dolphin 28	BL72
10	2	5	40	$4,200.00	Axxon II	Dolphin 40	NO27
11	2	6	40	$4,200.00	Karvel	Ray 4025	TR72

Alexamara provides boat maintenance service for owners at its two marinas. The types of service provided are stored in the SERVICE_CATEGORY table shown in Figure 1.11. A category number uniquely identifies each service that Alexamara performs. The table also contains a description of the category.

FIGURE 1.11 SERVICE_CATEGORY table

SERVICE_CATEGORY

CATEGORY_NUM	CATEGORY_DESCRIPTION
1	Routine engine maintenance
2	Engine repair
3	Air conditioning
4	Electrical systems
5	Fiberglass repair
6	Canvas installation
7	Canvas repair
8	Electronic systems (radar, GPS, autopilots, etc.)

Information about the services requested by owners is stored in the SERVICE_REQUEST table shown in Figure 1.12. Each row in the table contains a service ID that identifies each service request. The slip ID identifies the location (marina number and slip number) of the boat to be serviced. For example, the slip ID on the second row is 5. As indicated in the MARINA_SLIP table, the slip ID 5 identifies the boat in marina 1 and slip number B2. The SERVICE_REQUEST table also contains the category number of the service to be performed, a description of the specific service to be performed, and a description of the current status of the service. It also contains the estimated number of hours required to complete the service. For completed jobs, the table contains the actual number of hours it took to complete the service. If another appointment is required to complete additional service, the appointment date appears in the NEXT_SERVICE_DATE column.

FIGURE 1.12 SERVICE_REQUEST table

SERVICE_REQUEST

SERVICE_ ID	SLIP_ ID	CATEGORY_ NUM	DESCRIPTION	STATUS	EST_ HOURS	SPENT_ HOURS	NEXT_ SERVICE_ DATE
1	1	3	Air conditioner periodically stops with code indicating low coolant level. Diagnose and repair.	Technician has verified the problem. Air conditioning specialist has been called.	4	2	7/12/2007
2	5	4	Fuse on port motor blown on two occasions. Diagnose and repair.	Open	2	0	7/12/2007

(continued)

FIGURE 1.12   SERVICE_REQUEST table (continued)

SERVICE_REQUEST

SERVICE_ ID	SLIP_ ID	CATEGORY_ NUM	DESCRIPTION	STATUS	EST_ HOURS	SPENT_ HOURS	NEXT_ SERVICE_ DATE
3	4	1	Oil change and general routine maintenance (check fluid levels, clean sea strainers, etc.).	Service call has been scheduled.	1	0	7/16/2007
4	1	2	Engine oil level has been dropping drastically. Diagnose and repair.	Open	2	0	7/13/2007
5	3	5	Open pockets at base of two stantions.	Technician has completed the initial filling of the open pockets. Will complete the job after the initial fill has had sufficient time to dry.	4	2	7/13/2007
6	11	4	Electric-flush system periodically stops functioning. Diagnose and repair.	Open	3	0	
7	6	2	Engine overheating. Loss of coolant. Diagnose and repair.	Open	2	0	7/13/2007
8	6	2	Heat exchanger not operating correctly.	Technician has determined that the exchanger is faulty. New exchanger has been ordered.	4	1	7/17/2007
9	7	6	Canvas severely damaged in windstorm. Order and install new canvas.	Open	8	0	7/16/2007
10	2	8	Install new GPS and chart plotter.	Scheduled	7	0	7/17/2007

(continued)

FIGURE 1.12   SERVICE_REQUEST table (continued)

SERVICE_REQUEST

SERVICE_ ID	SLIP_ ID	CATEGORY_ NUM	DESCRIPTION	STATUS	EST_ HOURS	SPENT_ HOURS	NEXT_ SERVICE_ DATE
11	2	3	Air conditioning unit shuts down with HHH showing on the control panel.	Technician not able to replicate the problem. Air conditioning unit ran fine through multiple tests. Owner to notify technician if the problem recurs.	1	1	
12	4	8	Both speed and depth readings on data unit are significantly less than the owner thinks they should be.	Technician has scheduled appointment with owner to attempt to verify the problem.	2	0	7/16/2007
13	8	2	Customer describes engine as making a clattering sound.	Technician suspects problem with either propeller or shaft and has scheduled the boat to be pulled from the water for further investigation.	5	2	7/12/2007
14	7	5	Owner accident caused damage to forward portion of port side.	Technician has scheduled repair.	6	0	7/13/2007
15	11	7	Canvas leaks around zippers in heavy rain. Install overlap around zippers to prevent leaks.	Overlap has been created. Installation has been scheduled.	8	3	7/17/2007

■ SUMMARY

- Premiere Products is an organization whose information requirements include reps, customers, parts, orders, and order lines.

- Henry Books is an organization whose information requirements include

branches, publishers, authors, books, inventory, and author sequence.

- Alexamara Marina Group is an organization whose information requirements include marinas, owners, slips, service categories, and service requests.

■ KEY TERMS

database

■ EXERCISES (Premiere Products)

Answer each of the following questions using the Premiere Products data shown in Figure 1.2. No computer work is required.

1. List the names of all customers that have a credit limit of $7,500 or less.

2. List the order numbers for orders placed by customer number 608 on 10/23/2007.

3. List the part number, part description, and on-hand value for each part in item class SG. (*Hint:* On-hand value is the result of multiplying the number of units on hand by the price.)

4. List the part number and part description of all parts that are in item class HW.

5. How many customers have a balance that exceeds their credit limit?

6. What is the part number, description, and price for the least expensive part in the database?

7. For each order, list the order number, order date, customer number, and customer name.

8. For each order placed on October 21, 2007, list the order number, customer number, and customer name.

9. List the sales rep number and name for every sales rep who represents at least one customer with a credit limit of $10,000.

10. For each order placed on October 21, 2007, list the order number, part number, part description, and item class for each part ordered.

■EXERCISES (Henry Books)

Answer each of the following questions using the Henry Books data shown in Figures 1.4 through 1.7. No computer work is required.

1. List the name of each publisher that is located in New York.

2. List the name of each branch that has at least nine employees.

3. List the book code and title of each book that has the type FIC.

4. List the book code and title of each book that has the type FIC and that is in paperback.

5. List the book code and title of each book that has the type FIC or whose publisher code is SC.

6. List the book code and title of each book that has the type MYS and a price of less than $20.00.

7. Customers who are part of a special program get a 10% discount off regular book prices. For the first five books in the BOOK table, list the book code, title, and discounted price. (Use the PRICE column to calculate the discounted price.)

8. Find the name of each publisher containing the word "and."

9. List the book code and title of each book that has the type FIC, MYS, or ART.

10. How many books have the type SFI?

11. Calculate the average price for books that have the type ART.

12. For each book published by Penguin USA, list the book code and title.

13. List the book code, book title, and units on hand for each book in branch number 3.

■EXERCISES (Alexamara Marina Group)

Answer each of the following questions using the Alexamara Marina Group data shown in Figures 1.8 through 1.12. No computer work is required.

1. List the owner number, last name, and first name of every boat owner.

2. List the last name and first name of every owner located in Bowton.

3. List the marina number and slip number for every slip whose length is equal to or less than 30 feet.

4. List the marina number and slip number for every boat with the type Dolphin 28.

5. List the slip number for every boat with the type Dolphin 28 that is located in marina 1.

6. List the boat name for each boat located in a slip whose length is between 25 and 30 feet.

7. List the slip number for every slip in marina 1 whose annual rental fee is less than $3,000.00.

8. Labor is billed at the rate of $60.00 per hour. List the slip ID, category number, estimated hours, and estimated labor cost for every service request. To obtain the estimated labor cost, multiply the estimated hours by 60. Use the column name ESTIMATED_COST for the estimated labor cost.

9. List the marina number and slip number for all slips containing a boat with the type Sprite 4000, Sprite 3000, or Ray 4025.

10. How many Dolphin 25 boats are stored at both marinas?

11. For every boat, list the marina number, slip number, boat name, owner number, owner's first name, and owner's last name.

12. For every service request for routine engine maintenance, list the slip ID, the description, and the status.

13. For every service request for routine engine maintenance, list the slip ID, marina number, slip number, estimated hours, spent hours, owner number, and owner's last name.

CHAPTER 2

An Introduction to SQL

OBJECTIVES

- Understand the concepts and terminology associated with relational databases

- Create and run SQL commands in Oracle, Microsoft Access, and MySQL

- Create tables using SQL

- Identify and use data types to define columns in SQL tables

- Understand and use nulls

- Add rows to tables

- Describe a table's layout using SQL

Introduction

You already might be an experienced user of a database management system (DBMS). You might find a DBMS at your school's library, at a site on the Internet, or in any other place where you retrieve data using a computer. In this chapter, you will learn about the concepts and terminology associated with the relational model for database management. Then you will learn how to create a database by describing and defining the tables and columns that make up the database. In this text, you will study a language called **SQL (Structured Query Language)**, which is the one of the most popular and widely used languages for retrieving and manipulating database data.

In the mid-1970s, SQL was developed as the data manipulation language for IBM's prototype relational model DBMS, System R, under the name SEQUEL at IBM's San Jose research facilities. In 1980, the language was renamed SQL (but still pronounced "sequel" although the equally popular pronunciation of "S-Q-L" ["ess-cue-ell"] is used in this book) to avoid confusion with an unrelated hardware product named SEQUEL. Most DBMSs use a version of SQL as their data manipulation language.

In this chapter, you will learn how to assign data types to columns in a database. You will also learn about a special type of value, called a null value, and learn how to handle these values during database creation. You will learn how to load a database by creating tables and adding data to them. Finally, you will learn how to describe a table's layout using SQL.

Note: This text demonstrates the use of four SQL implementations: Oracle 10*g* SQL*Plus, Oracle 10*g* SQL*Plus Worksheet, Microsoft Access 2003, and MySQL. However, the SQL commands used in this text should work for other SQL implementations with little or no modification.

■ Relational Databases

A relational database is essentially a collection of tables like the ones you viewed for Premiere Products in Chapter 1 and that also appear in Figure 2.1. You might wonder why a relational database isn't called a "table" database, or something similar, if a database is really just a collection of tables. Formally, these tables are called relations, and this is where a relational database gets its name.

FIGURE 2.1 Sample data for Premiere Products

REP

REP_NUM	LAST_NAME	FIRST_NAME	STREET	CITY	STATE	ZIP	COMMISSION	RATE
20	Kaiser	Valerie	624 Randall	Grove	FL	33321	$20,542.50	0.05
35	Hull	Richard	532 Jackson	Sheldon	FL	33553	$39,216.00	0.07
65	Perez	Juan	1626 Taylor	Fillmore	FL	33336	$23,487.00	0.05

CUSTOMER

CUSTOMER_NUM	CUSTOMER_NAME	STREET	CITY	STATE	ZIP	BALANCE	CREDIT_LIMIT	REP_NUM
148	Al's Appliance and Sport	2837 Greenway	Fillmore	FL	33336	$6,550.00	$7,500.00	20
282	Brookings Direct	3827 Devon	Grove	FL	33321	$431.50	$10,000.00	35
356	Ferguson's	382 Wildwood	Northfield	FL	33146	$5,785.00	$7,500.00	65
408	The Everything Shop	1828 Raven	Crystal	FL	33503	$5,285.25	$5,000.00	35
462	Bargains Galore	3829 Central	Grove	FL	33321	$3,412.00	$10,000.00	65
524	Kline's	838 Ridgeland	Fillmore	FL	33336	$12,762.00	$15,000.00	20
608	Johnson's Department Store	372 Oxford	Sheldon	FL	33553	$2,106.00	$10,000.00	65
687	Lee's Sport and Appliance	282 Evergreen	Altonville	FL	32543	$2,851.00	$5,000.00	35
725	Deerfield's Four Seasons	282 Columbia	Sheldon	FL	33553	$248.00	$7,500.00	35
842	All Season	28 Lakeview	Grove	FL	33321	$8,221.00	$7,500.00	20

ORDERS

ORDER_NUM	ORDER_DATE	CUSTOMER_NUM
21608	10/20/2007	148
21610	10/20/2007	356
21613	10/21/2007	408
21614	10/21/2007	282
21617	10/23/2007	608
21619	10/23/2007	148
21623	10/23/2007	608

ORDER_LINE

ORDER_NUM	PART_NUM	NUM_ORDERED	QUOTED_PRICE
21608	AT94	11	$21.95
21610	DR93	1	$495.00
21610	DW11	1	$399.99
21613	KL62	4	$329.95
21614	KT03	2	$595.00
21617	BV06	2	$794.95
21617	CD52	4	$150.00
21619	DR93	1	$495.00
21623	KV29	2	$1,290.00

FIGURE 2.1   Sample data for Premiere Products (continued)

PART

PART_NUM	DESCRIPTION	ON_HAND	CLASS	WAREHOUSE	PRICE
AT94	Iron	50	HW	3	$24.95
BV06	Home Gym	45	SG	2	$794.95
CD52	Microwave Oven	32	AP	1	$165.00
DL71	Cordless Drill	21	HW	3	$129.95
DR93	Gas Range	8	AP	2	$495.00
DW11	Washer	12	AP	3	$399.99
FD21	Stand Mixer	22	HW	3	$159.95
KL62	Dryer	12	AP	1	$349.95
KT03	Dishwasher	8	AP	3	$595.00
KV29	Treadmill	9	SG	2	$1,390.00

Entities, Attributes, and Relationships

The terms *entity*, *attribute*, and *relationship* are very important for you to know when discussing databases. An **entity** is a person, place, object, event, or idea for which you want to store and process data. The entities of interest to Premiere Products, for example, are customers, orders, parts, and sales reps. The entities of interest to a school include students, faculty, and classes; a real estate agency is interested in clients, houses, and agents; and a used car dealer is interested in vehicles, customers, and manufacturers.

An **attribute** is a characteristic or property of an entity. The term is used in this text very much as it is in everyday English. For the entity "person," for example, the list of attributes might include such things as eye color and height. For Premiere Products, the attributes of interest for the entity "customer" are such things as customer name, street, city, and so on. An attribute is also called a **field** or **column** in many database systems.

The final key term is relationship. A **relationship** is the association between entities. There is an association between customers and sales reps, for example, at Premiere Products. A sales rep is associated with all of his or her customers, and a customer is associated with its sales rep. Technically, you say that a sales rep is *related to* all of his or her customers, and a customer is *related to* its sales rep. This particular relationship is called a **one-to-many relationship** because each sales rep is associated with *many* customers, but each customer is associated with only *one* sales rep. In this type of relationship, the word *many* is used differently than in everyday English; it might not always indicate a large number. In this context, for example, the term *many* means that a sales rep can be associated with *any* number of customers. That is, one sales rep can be associated with zero, one, or more customers.

How does a DBMS that follows the relational model handle entities, attributes of entities, and relationships between entities? Entities and attributes are fairly simple. Each entity has its own table. In the Premiere Products database, there is one table for sales reps, one table for customers, and so on. The attributes of an entity become the columns in the table. In the table for sales reps, for example, there is a column for the sales rep number, a column for the sales rep last name, and so on.

What about relationships? At Premiere Products, there is a one-to-many relationship between sales reps and customers (each sales rep is related to the *many* customers that he or she represents, and each customer is related to the *one* sales rep who represents the customer). In a relational model database this relationship is represented by using common columns in two or more tables. Consider Figure 2.1 again. The REP_NUM column in the REP table and the REP_NUM column in the CUSTOMER table are used to implement the relationship between sales reps and customers. Given a sales rep, you can use these columns to identify all the customers that he or she represents; given a customer, you can use these columns to find the sales rep who represents the customer.

More formally, a relation is essentially a two-dimensional table. If you consider the tables shown in Figure 2.1, however, you might see that there are certain restrictions placed on relations. Each column in a table should have a unique name, and entries within each column should all "match" this column name. For example, in the CREDIT_LIMIT column, all entries should in fact *be* credit limits. Also, each row (also called a **record** or a **tuple** in some programs) should be unique. After all, if two rows in a table contain identical data, the second row doesn't provide any information that you don't already have. In addition, for maximum flexibility in manipulating data, the order in which columns and rows appear in a table should be immaterial. Finally, a table's design should be as simple as possible; you should restrict each position in a table to a single entry by not allowing multiple entries (called a **repeating group**) in an individual location in the table, as shown in Figure 2.2.

FIGURE 2.2   Poor table structure

ORDERS

ORDER_NUM	ORDER_DATE	CUSTOMER_NUM	PART_NUM	NUM_ORDERED	QUOTED_PRICE
21608	10/20/2007	148	AT94	11	$21.95
21610	10/20/2007	356	DR93	1	$495.00
			DW11	1	$399.99
21613	10/21/2007	408	KL62	4	$329.95
21614	10/21/2007	282	KT03	2	$595.00
21617	10/23/2007	608	BV06	2	$794.95
			CD52	4	$150.00
21619	10/23/2007	148	DR93	1	$495.00
21623	10/23/2007	608	KV29	2	$1,290.00

These ideas lead to the following definitions.

Definition: A **relation** is a two-dimensional table in which:

1. The entries in the table are single-valued; that is, each location in the table contains a single entry.

2. Each column has a distinct name (technically called the attribute name).

3. All values in a column are values of the same attribute (that is, all entries must match the column name).

4. The order of columns is immaterial.

5. Each row is distinct.

6. The order of rows is immaterial.

Definition: A **relational database** is a collection of relations.

There is a commonly accepted shorthand representation to show the structure of a relational database: After the name of the table, all the columns in the table are listed within a set of parentheses. In addition, each table should appear on its own line. Using this method, you would write the Premiere Products database as follows:

```
REP (REP_NUM, LAST_NAME, FIRST_NAME, STREET, CITY, STATE, ZIP,
    COMMISSION, RATE)
CUSTOMER (CUSTOMER_NUM, CUSTOMER_NAME, STREET, CITY, STATE, ZIP,
    BALANCE, CREDIT_LIMIT, REP_NUM)
ORDERS (ORDER_NUM, ORDER_DATE, CUSTOMER_NUM)
ORDER_LINE (ORDER_NUM, PART_NUM, NUM_ORDERED, QUOTED_PRICE)
PART (PART_NUM, DESCRIPTION, ON_HAND, CLASS, WAREHOUSE, PRICE)
```

Note: In general, SQL is not case-sensitive; you can type commands using uppercase or lowercase letters. There is one exception to this rule, however. When you are inserting character values into a table, you must use the correct case.

The Premiere Products database contains some duplicate column names. For example, the REP_NUM column appears in both the REP and CUSTOMER tables. Suppose a situation exists in which the two columns might be confused. If you write REP_NUM, how would the DBMS or another programmer know which REP_NUM column you intend to use? You need a way to associate the correct table with the column name. One common approach to this problem is to write both the table name and the column name, separated by a period. Thus, you would write the REP_NUM column in the CUSTOMER table as CUSTOMER.REP_NUM and the REP_NUM column in the REP table as REP.REP_NUM. Technically, when you combine a column name with a table name, you say that you **qualify** the column name. It is always acceptable to qualify column names, even if there is no possibility of confusion. If confusion is possible, of course, it is essential to qualify column names.

The **primary key** of a table (relation) is the column or collection of columns that uniquely identifies a given row in that table. In the REP table, the sales rep number uniquely identifies a given row. For example, sales rep 35 occurs in only one row of the table; the REP_NUM column is the table's primary key. You usually indicate a table's primary key by underlining the column or collection of columns that comprises the primary key for each table in the database. The complete representation for the Premiere Products database, where the underlined column name(s) indicates each table's primary key, is as follows:

```
REP (REP_NUM, LAST_NAME, FIRST_NAME, STREET, CITY, STATE, ZIP,
     COMMISSION, RATE)
CUSTOMER (CUSTOMER_NUM, CUSTOMER_NAME, STREET, CITY, STATE, ZIP,
     BALANCE, CREDIT_LIMIT, REP_NUM)
ORDERS (ORDER_NUM, ORDER_DATE, CUSTOMER_NUM)
ORDER_LINE (ORDER_NUM, PART_NUM, NUM_ORDERED, QUOTED_PRICE)
PART (PART_NUM, DESCRIPTION, ON_HAND, CLASS, WAREHOUSE, PRICE)
```

QUESTION Why does the primary key of the ORDER_LINE table consist of two columns, instead of just one?

ANSWER No single column in the ORDER_LINE table uniquely identifies a given row in the ORDER_LINE table. The ORDER_NUM column cannot be the primary key; there are two rows with order number 21610, for example. Likewise, the PART_NUM column cannot be the primary key; there are two rows with part number DR93. To uniquely identify a row requires the combination of two columns: ORDER_NUM and PART_NUM.

■ Database Creation

Before you begin loading and accessing data in a table, you must describe the layout of each table that the database will contain.

EXAMPLE 1 : Describe the layout of the REP table to the DBMS.

You use the **CREATE TABLE** command to describe the layout of a table. The word TABLE is followed by the name of the table to be created and then by the names and data types of the columns that will comprise the table. The **data type** indicates the type of data that the column can contain (for example, characters, numbers, or dates) as well as the maximum number of characters or digits that the column can store. The rules for naming tables and columns vary slightly from one implementation of SQL to another. If you have any questions about the validity of any of the names you have chosen for tables or columns, consult your system's manual or its online Help system.

Some common restrictions placed on table and column names are:

- The names cannot exceed 18 characters. (In Oracle, names can be up to 30 characters in length.)

- The names must start with a letter.

- The names can contain letters, numbers, and underscores (_).

- The names cannot contain spaces.

The names used in this text should work for any SQL implementation. The appropriate SQL command for Example 1 is shown in Figure 2.3.

FIGURE 2.3
• • • • • • • • • • • • • •
CREATE TABLE
command for the
REP table

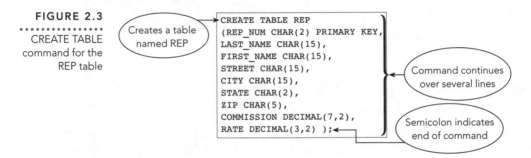

This CREATE TABLE command, which uses the data definition features of SQL, describes a table named REP. The table contains nine columns: REP_NUM, LAST_NAME, FIRST_NAME, STREET, CITY, STATE, ZIP, COMMISSION, and RATE. The REP_NUM column can store two characters and is the primary key. The LAST_NAME column can store 15 characters, and the STATE column can store two characters. The COMMISSION column can store only numbers and those numbers are limited to seven digits, including two decimal places. Similarly, the RATE column can store three numbers, including two decimal places. You can think of the SQL command in Figure 2.3 as creating an empty table with column headings for each column name.

In SQL, commands are free-format; that is, no rule says that a particular word must begin in a particular position on the line. For example, you could have written the CREATE TABLE command shown in Figure 2.3 as follows:

```
CREATE TABLE REP (REP_NUM CHAR(2) PRIMARY KEY, LAST_NAME CHAR(15),
FIRST_NAME CHAR(15), STREET CHAR(15), CITY CHAR(15), STATE CHAR(2),
ZIP CHAR(5), COMMISSION DECIMAL(7,2), RATE DECIMAL(3,2) );
```

The manner in which the CREATE TABLE command shown in Figure 2.3 was written simply makes the command more readable. This text will strive for such readability when writing SQL commands.

You can press the Enter key to move to the next line in a command. You indicate the end of a command by typing a semicolon.

■ Running SQL Commands

The precise manner in which you run SQL commands depends on the program in which you are working and, in some cases, your individual preferences. If you are using Oracle 10*g*, you can complete your work in SQL*Plus or SQL*Plus Worksheet.

SQL*PLUS : **Oracle SQL*Plus** is a program in which you type SQL commands at an
USER : SQL> prompt. You press the Enter key at the end of each line. To end and
: execute a command, type a semicolon and press the Enter key. For exam-
: ple, the command shown in Figure 2.4 creates the REP table.

FIGURE 2.4   Using Oracle SQL*Plus to create a table

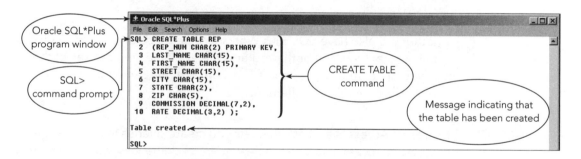

SQL*PLUS : **Oracle SQL*Plus Worksheet** is a GUI (graphical user interface) program in
WORKSHEET : which you type commands in the upper pane of the window and then click
USER : the Execute button to run the command. Oracle executes the command
: and displays the result in the lower pane of the window. For example, the
: command shown in Figure 2.5 creates the REP table.

FIGURE 2.5 Using Oracle SQL*Plus Worksheet to create a table

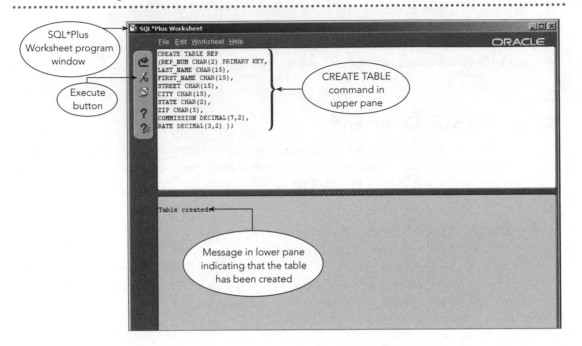

If you are using Oracle, you probably have access to SQL*Plus. However, whether you also have SQL*Plus Worksheet depends on your particular installation of Oracle. If both programs are available, you can select the one you prefer (they both generally accomplish the same tasks) unless your instructor indicates differently.

A CCESS
USER

Microsoft Access is a DBMS that lets you create queries in SQL view. To use SQL in Access, you create a new query, click the View button list arrow on the toolbar, and then click SQL View. Then you can type an SQL command in the window that opens and click the Run button on the toolbar to execute the command. For example, the command shown in Figure 2.6 creates the REP table. If you click the Run button, Access will create the table shown in the CREATE TABLE command. (Access doesn't display a message indicating the result was successful.)

FIGURE 2.6   Using Access SQL view to create a table

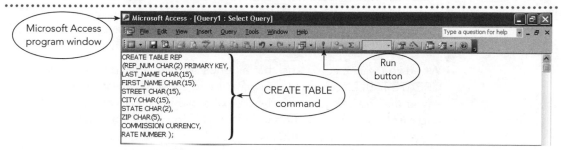

Access doesn't have a DECIMAL data type. To create numbers with deci-
mals, you must use either the CURRENCY or NUMBER data type. Use the
CURRENCY data type for fields that will contain currency values; use the
NUMBER data type for all other numeric fields.

In Access, it is common to create a database using Table Design view and
then add records using either Datasheet or Form view. You can still use
tables created this way in the SQL commands you will encounter in the
remainder of this text.

MYSQL
USER

MySQL is an open-source database system that supports the SQL language. In
MySQL, you type commands at the mysql> prompt, as shown in Figure 2.7.
The first command in the figure (use premiere;) activates the database named
premiere (that is, the Premiere Products database). After you execute this command, all subsequent commands during the current session will affect the
Premiere Products database. (You do not have to execute this command again
during the current session unless you need to change to another database.)

FIGURE 2.7   Using MySQL to create a table

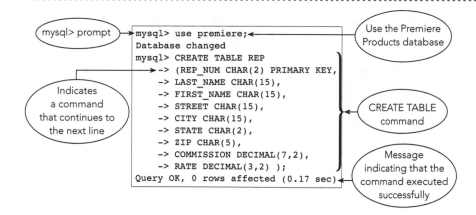

These commands assume that the database already exists. If it did not, you would need one additional command (create premiere;) to create the Premiere Products database. Then you would use the use premiere; command to access the Premiere Products database.

When you type a command that extends over several lines, such as the CREATE TABLE command shown in Figure 2.7, you can press the Enter key at the end of each line. MySQL will move the cursor to the next line and display the continuation indicator (->). To finish a command, type a semicolon at the end of the last line of the command and then press the Enter key. MySQL will execute the command and indicate whether the command was successful, as shown in Figure 2.7.

Editing SQL Commands

When you use Oracle SQL*Plus Worksheet or Access SQL view to complete your work in SQL, you can edit your commands to change them or correct errors by using the mouse and the correction techniques you might use in a word processor. For example, you can use the mouse to select text and type to replace it or click the insertion point to position it on a line. After making your edits, you can click the Execute button (SQL*Plus Worksheet) or the Run button (Access) to execute the command. In Oracle SQL*Plus and MySQL, however, you must edit the on-screen command to make changes to it.

SQL*PLUS USER : You cannot edit entire commands in Oracle SQL*Plus; you must edit them one line at a time. In SQL*Plus, the most recent command you entered is stored in a special area of memory called the **command buffer** (also called the **buffer**). You can edit the command in the buffer by using the editing commands shown in Figure 2.8.

FIGURE 2.8   Oracle SQL*Plus editing commands

Activity	Command	Abbreviation
Add text at end of current line	APPEND text	A text
Change current line	Type the line number	
Change text in current line	CHANGE /old/new	C /old/new
Delete all lines from buffer	CLEAR BUFFER	CL BUFF
Delete current line	DEL	
Edit the entire command currently in the buffer using an editor such as Notepad	EDIT	
Insert a line following current line	INPUT	I
List the command currently in the buffer	LIST	L
Run the command currently in the buffer	RUN	R

: Consider the SQL command shown in Figure 2.9.

FIGURE 2.9   CREATE TABLE command with errors (Oracle SQL*Plus)

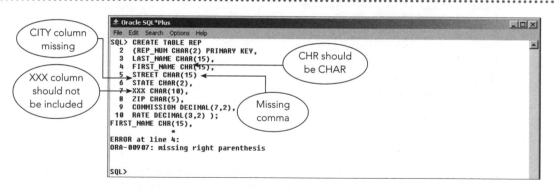

There are several mistakes in the command shown in Figure 2.9. In line 4, CHAR is misspelled. Line 5 is missing a comma. The CITY column is missing, and line 7 contains errors and should be deleted.

Figure 2.10 illustrates how you can use the editing commands in Oracle SQL*Plus to make the necessary corrections. The first command, L, lists the command in the buffer. The second command, 4, moves the current position to line 4 in the buffer. The C /CHR/CHAR command changes the misspelled CHR to CHAR. The next two commands move the current position to line 5 and add a comma at the end of the line. The 7 and DEL commands move the current position to line 7 and then delete the line. Finally, the 5 and INPUT commands move the current position to line 5 and then insert a new line following the current line 5. After inserting this line, you can also insert additional lines. To indicate that you are finished adding lines, you would press the Enter key.

FIGURE 2.10

Correcting errors in a command (Oracle SQL*Plus)

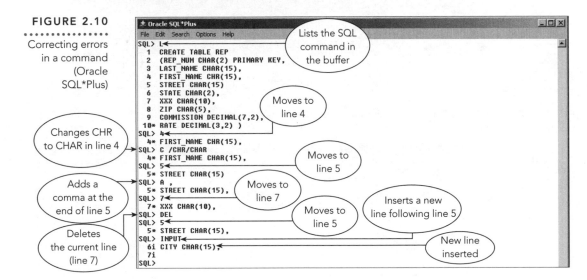

After making changes, it's a good idea to use the L command to list the modified SQL command to verify that your changes are correct. If you type a semicolon after a command, Oracle will run (execute) the command immediately. To run the command you edited, type RUN. Typing RUN displays the command again before it is executed. If you simply want to execute the command without first displaying it, type RUN followed by a slash (/).

MYSQL USER : In MySQL, the most recent command you entered is stored in a special area of memory called the **statement history**. You can edit the command in the statement history by using the editing commands shown in Figure 2.11.

FIGURE 2.11 MySQL editing commands

Activity	Editing Key
Move up a line in the statement history	Up arrow
Move down a line in the statement history	Down arrow
Move left one character on the current line	Left arrow
Move right one character on the current line	Right arrow
Move to the beginning of the current line	Ctrl + A
Move to the end of the current line	Ctrl + E
Delete the previous character	Backspace
Delete the character above the cursor	Delete

There are several mistakes in the command shown in Figure 2.12. In line 4, CHAR is misspelled. Line 5 is missing a comma. The CITY column is missing, and line 7 contains errors and should be deleted.

FIGURE 2.12 CREATE TABLE command with errors (MySQL)

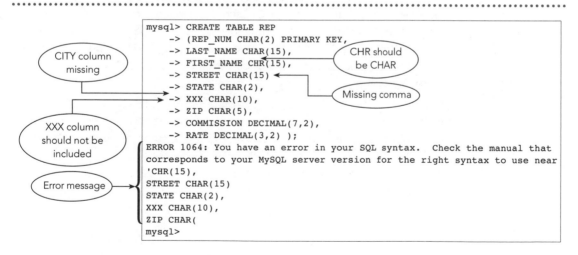

To begin the editing process, press the Up arrow key to move back through the lines in the command until the first line, CREATE TABLE REP, appears after the mysql> prompt. Because this line is correct, press the Enter key. MySQL will move the cursor to the second line and display the continuation indicator (->). Use the Down arrow key to move to the second line, containing the REP_NUM column. Because this line is also correct, press the Enter key. Use the Down arrow key to move to the third line, containing the LAST_NAME column, and then press the Enter key to indicate that it is correct. Move to the fourth line, containing the FIRST_NAME column; this line contains an error. Use the Left or Right arrow key as necessary to move the cursor to the R in CHR, and type the letter A to change CHR to CHAR. (If the A replaced the R so that you changed CHR to CHA, you were not in Insert mode. Press the Insert key and then type the letter R.) When the line is correct, press the Enter key.

Move to the line containing the STREET column, type a comma at the end of the line, and then press the Enter key. You need to insert the line for the CITY column after the line containing the STREET column. Rather than moving to another line in the statement history, simply type the correct line (CITY CHAR(15),) and then press the Enter key. Move to the line containing the STATE column and press the Enter key. Move to the line containing the ZIP column and press the Enter key. In the process, you will move right through the line containing the XXX error. Because you did not press the Enter key when that line was on the screen, it will not be part of the revised command. Move to the line containing the COMMISSION column and press the Enter key. Move to the line containing the RATE column and press the Enter key. Because this line ends with a semicolon, MySQL will execute the revised command.

■ Dropping a Table

Another way of correcting errors in a table is to delete (drop) it and start over. For example, suppose your CREATE TABLE command contains a column named LST instead of LAST or defines a column as CHAR(5) instead of CHAR(15). Suppose you don't discover the error and you execute the command, creating a table with errors. In this case, you can delete the entire table using the **DROP TABLE** command and then create the correct table using the appropriate CREATE TABLE command. The command DROP TABLE is followed by the name of the table you want to delete and a semicolon. To delete the REP table, for example, you would enter the following command:

```
DROP TABLE REP;
```

You should note that dropping a table also deletes any data that you entered into the table. It's a good idea to check your CREATE TABLE commands carefully before executing them and to correct any problems before adding data. Later in this text, you will learn how to change a table's structure without having to delete the entire table.

Q**UESTION**   How can I correct a mistake that I made when I created a table?

ANSWER   Later in the text, you will see how to alter your table to make any necessary corrections. For now, the easiest way is to drop the table using the DROP TABLE command and then execute the correct CREATE TABLE command.

■ Data Types

For each column in a table, you must specify the type of data that the column will store. Although the actual data types vary slightly from one SQL implementation to another, Figure 2.13 indicates some common data types.

FIGURE 2.13 Common data types

Data Type	Description
CHAR(*n*)	Stores a character string *n* characters long. You use the CHAR type for columns that contain letters and special characters and for columns containing numbers that will not be used in any arithmetic operations. Because neither sales rep numbers nor customer numbers will be used in any arithmetic operations, for example, the REP_NUM and CUSTOMER_NUM columns are both assigned the CHAR data type.
DATE	Stores date data. The specific format in which dates are stored varies from one SQL implementation to another. In Oracle, dates are enclosed in single quotation marks and have the form DD-MON-YYYY (for example, '15-OCT-2007' is October 15, 2007). In Access, dates are enclosed in number signs and have the form MM/DD/YYYY (for example, #10/15/2007# is October 15, 2007). In MySQL, dates are enclosed in single quotation marks and have the form YYYY-MM-DD (for example, '2007-10-15' is October 15, 2007).
DECIMAL(*p*,*q*)	Stores a decimal number *p* digits long with *q* of these digits being decimal places to the right of the decimal point. For example, the data type DECIMAL(5,2) represents a number with three places to the left and two places to the right of the decimal (for example, 100.00). You can use the contents of DECIMAL columns for arithmetic. (*Note:* The specific meaning of DECIMAL varies from one SQL implementation to another. In some implementations, the decimal point counts as one of the places, and in other implementations it does not. Likewise, in some implementations a minus sign counts as one of the places, but in others it does not.)
INTEGER	Stores integers, which are numbers without a decimal part. The valid range is –2147483648 to 2147483647. You can use the contents of INTEGER columns for arithmetic.
SMALLINT	Stores integers, but uses less space than the INTEGER data type. The valid range is –32768 to 32767. SMALLINT is a better choice than INTEGER when you are certain that the column will store numbers within the indicated range. You can use the contents of SMALLINT columns for arithmetic.

■ Nulls

Occasionally, when you enter a new row into a table or modify an existing row, the values for one or more columns are unknown or unavailable. For example, you can add a customer's name and address to a table even though the customer does not have an assigned sales rep or an established credit limit. In other cases, some values might never be known; perhaps there is a customer that does not have a sales rep. In SQL, you handle this problem by using a special value to represent situations in which an actual value is unknown, unavailable, or not applicable. This special value is called a **null data value**, or simply a **null**. When creating a table, you can specify whether to allow nulls in the individual columns.

QUESTION   Should a user be allowed to enter null values for the primary key?

ANSWER   No. The primary key is supposed to uniquely identify a given row, and this would be impossible if nulls were allowed. For example, if you stored two customer records without values in the primary key column, you would have no way to tell them apart.

Implementation of Nulls

In SQL, you use the **NOT NULL** clause in a CREATE TABLE command to indicate columns that *cannot* contain null values. The default is to allow nulls; columns for which you do not specify NOT NULL can accept null values.

For example, suppose that the LAST_NAME and FIRST_NAME columns in the REP table cannot accept null values, but all other columns in the REP table can. The following CREATE TABLE command would accomplish this goal:

```
CREATE TABLE REP
(REP_NUM CHAR(2) PRIMARY KEY,
LAST_NAME CHAR(15) NOT NULL,
FIRST_NAME CHAR(15) NOT NULL,
STREET CHAR(15),
CITY CHAR(15),
STATE CHAR(2),
ZIP CHAR(5),
COMMISSION DECIMAL(7,2),
RATE DECIMAL(3,2) );
```

The system will reject any attempt to store a null value in either the LAST_NAME or FIRST_NAME column. The system will accept an attempt to store a null value in the STREET column, however, because the STREET column can accept null values. Because the primary key column cannot accept null values, you don't need to specify the REP_NUM column as NOT NULL.

■ Loading a Table with Data

After you have created the tables in a database, you can load data into them by using the INSERT command.

The INSERT Command

The **INSERT** command adds rows to a table. You type INSERT INTO followed by the name of the table into which you are adding data. Then you type the VALUES command followed by the specific values to be inserted in parentheses. When adding rows to character columns, make sure you enclose the values in single quotation marks (for example, 'Kaiser'). You must also enter the values in the appropriate case, because character data is stored exactly as you enter it.

Note: You must enclose values in single quotation marks for any column whose type is character (CHAR) even if the data contains numbers. Because the ZIP column in the REP table has a CHAR data type, for example, you must enclose zip codes in single quotation marks, even though they are numbers.

Note: If you need to enter an apostrophe (single quotation mark), you type two single quotation marks. For example, to enter the name O'Toole, you would type 'O''Toole' as the value in the INSERT command.

EXAMPLE 2 : Add sales rep 20 from Figure 2.1 to the database.

The command for this example is shown in Figure 2.14. Note that the character strings ('20','Kaiser','Valerie', and so on) are enclosed in single quotation marks. After you execute the command, the first record will be added to the REP table.

FIGURE 2.14
. . . .
INSERT command
for the first record
in the REP table

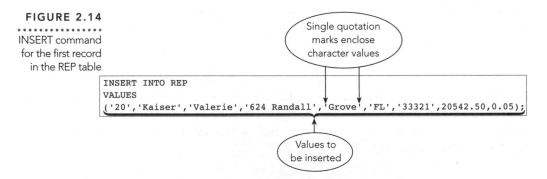

Note: Make sure that you type the values in the same case as those shown in the figures to avoid problems later when retrieving data from the database.

EXAMPLE 3 : Add sales reps 35 and 65 from Figure 2.1 to the REP table.

You could enter and execute new INSERT commands to add the new rows to the table. However, an easier, faster way is to modify the previous INSERT command and execute it to add the record for the second sales rep and then modify and execute the INSERT command again for the third sales rep. In Oracle SQL*Plus Worksheet and Access SQL view, you can simply select text, make the desired changes in the command on the screen, and then execute the new command. If you are using Oracle SQL*Plus or MySQL, however, you have to edit the commands manually.

SQL*PLUS : In Oracle SQL*Plus you must start a text editor program to edit a com-
USER : mand. First you execute the L command to list the command currently
: stored in the buffer. In this case, it is the previous INSERT command. Then
: you type the word EDIT, as shown in Figure 2.15. When you press the
: Enter key, you will execute the EDIT command.

FIGURE 2.15 Modifying the INSERT command (Oracle SQL*Plus)

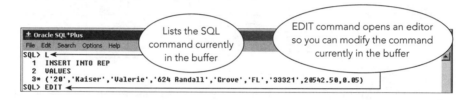

: When you execute the EDIT command, Oracle SQL*Plus opens a text edi-
: tor, such as Notepad, which you can use to update the command stored in
: the buffer. Figure 2.16 shows the previous INSERT command from the
: buffer in the Notepad program window. You can edit this command like
: you would text in a word processor.

FIGURE 2.16 Using an editor to modify the INSERT command (Oracle SQL*Plus)

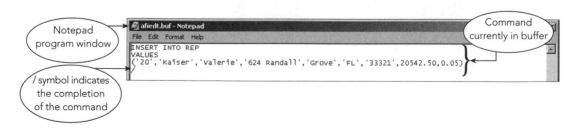

Oracle inserted the slash (/) at the end of the command to indicate that the command was complete. When you finish editing the command and save it, the slash indicates to Oracle that the command is complete and Oracle displays an SQL> prompt. If you omit the slash and save and close Notepad, Oracle assumes you are still working on the command and displays the next line number in the command, rather than the SQL> prompt. You don't need to type a semicolon to end your commands when you edit them.

Figure 2.17 shows the edited command. Now you can save the changes, close the editor, and type RUN at the SQL> prompt to execute the modified command. Oracle SQL*Plus lists the new command and then displays a message that one row was created in the table.

FIGURE 2.17   Modified INSERT command (Oracle SQL*Plus)

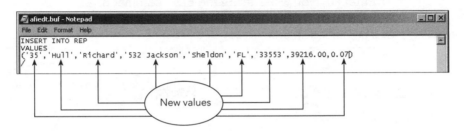

Figure 2.18 shows the results of using this technique to add all three records to the REP table in Oracle SQL*Plus.

FIGURE 2.18   Inserting additional rows (Oracle SQL*Plus)

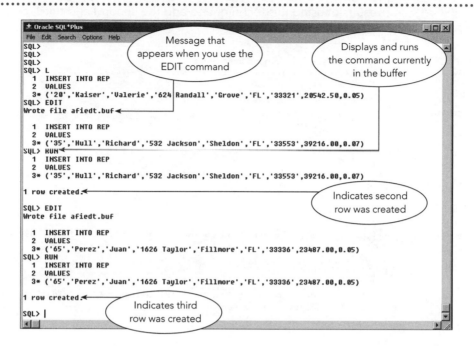

MYSQL USER : To edit a command in MySQL, you can use the same techniques you used to edit a command to add the two additional rows instead of typing two new INSERT commands. You can use the Up arrow key to bring the line from the previous INSERT command to the mysql> prompt and then press the Enter key. Use the Down arrow key to bring the VALUES line in the command to the screen and then press the Enter key. Finally, use the Down arrow key to bring the line containing the values from the previous INSERT command to the screen, change the values to the values for the row you wish to add, and then press the Enter key. Figure 2.19 shows the results of using this technique to add the second and third rows to the REP table.

FIGURE 2.19 Modifying the INSERT command in MySQL

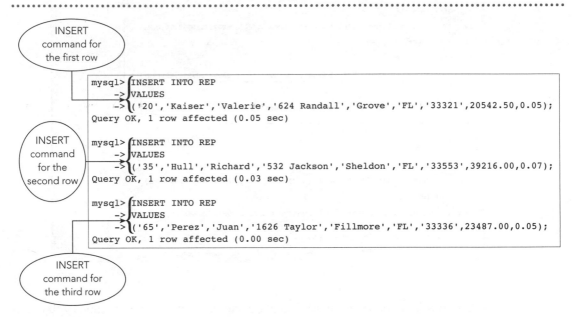

The INSERT Command with Nulls

To enter a null value into a table, you use a special format of the INSERT command. In this special format, you identify the names of the columns that will accept non-null values, and then list only these non-null values after the VALUES command, as shown in Example 4.

EXAMPLE 4 : Add sales rep 85 to the REP table. Her name is Tina Webb. All columns except REP_NUM, LAST_NAME, and FIRST_NAME are null.

In this case you do not enter a value of null; you enter only the non-null values. To do so, you must indicate precisely which values you are entering by listing the corresponding columns as shown in Figure 2.20. The command shown in the figure indicates that you are entering data in only the REP_NUM, LAST_NAME, and FIRST_NAME columns and that you *will not* enter values in any other columns.

FIGURE 2.20

· · · · · · · · · · · · · · · ·
Inserting a row
containing null
values

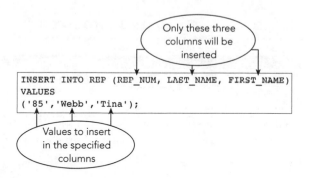

■ Viewing Table Data

To view the data in a table, you use the **SELECT** command, which is complex and the subject of the next two chapters. There is a simple version, however, that you can use to display all the rows and columns in a table: It consists of the word SELECT, followed by an asterisk, followed by the word FROM and then the name of the table containing the data you want to view. Just as with other SQL commands, the command ends with a semicolon.

SQL*PLUS : In SQL*Plus, you type the command at the SQL> prompt and press the
    USER : Enter key at the end of each line. When you type a semicolon at the end
         : of the last line of the command and press the Enter key, Oracle will exe-
         : cute the query and display its results, as shown in Figure 2.21. Notice that
         : the STATE column heading was truncated to fit the data.

FIGURE 2.21   Using a SELECT command to view table data in Oracle SQL*Plus

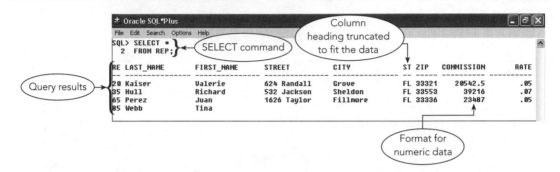

You might need to adjust the number of characters displayed to ensure
that the data fits on the screen. You can adjust the line size using the SET
LINESIZE command. To change the line size to 100 characters per line,
for example, you would execute the following command before running
the query:

```
SET LINESIZE 100
```

After you have executed the SET LINESIZE command, the change will
remain in effect throughout your entire session.

SQL*PLUS : In SQL*Plus Worksheet, you type a command in the upper pane and then
WORKSHEET : click the Execute button. Oracle will then execute the query and display its
    USER : results in the lower pane as shown in Figure 2.22. Notice that the STATE
         : column heading was truncated to fit the data.

FIGURE 2.22 Using a SELECT command to view table data in Oracle SQL*Plus Worksheet

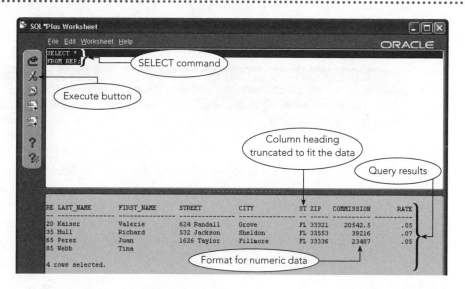

You might need to adjust the number of characters displayed to ensure that the data fits on the screen. You can adjust the line size using the SET LINESIZE command. To change the line size to 100 characters per line, for example, you would execute the following command before running the query:

`SET LINESIZE 100`

After you have executed the SET LINESIZE command, the change will remain in effect throughout your entire session.

In Access, type the query in SQL view as shown in Figure 2.23.

FIGURE 2.23 Using a SELECT command to view table data in Access

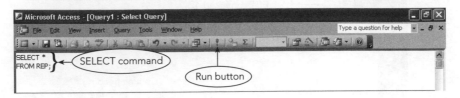

When the query is complete, click the Run button. Access will run the query and display its results in Datasheet view as shown in Figure 2.24. If the data displayed does not fit on the screen, you can adjust the columns to best fit the data they contain by double-clicking the right edge of each column heading. You can use the scroll bars to view data that has scrolled off the screen.

FIGURE 2.24 Query results in Access Datasheet view

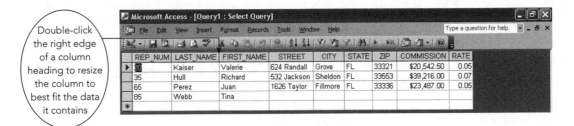

In MySQL, you type the command at the mysql> prompt and press the Enter key at the end of each line. When you type the semicolon at the end of the last line and press the Enter key, MySQL will execute the query and display its results as shown in Figure 2.25.

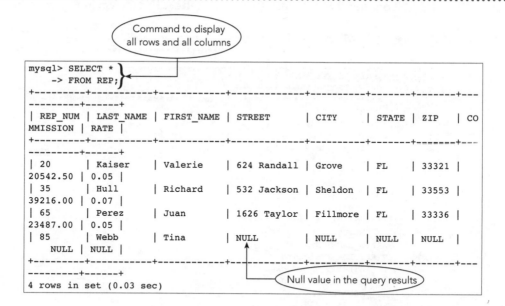

The query results contain line breaks because the data is too wide to fit on the screen, making the output difficult to read. In many cases, the query results will fit on the screen without any adjustments. In cases where the data will not fit on the screen, you can display the data vertically rather than horizontally. To do so, replace the semicolon at the end of the command with a slash (\) and the letter G as shown in Figure 2.26.

FIGURE 2.26   Using a SELECT command to view table data vertically in MySQL

```
mysql> SELECT *
    -> FROM REP\G
*************************** 1. row ***************************
    REP_NUM: 20
  LAST_NAME: Kaiser
 FIRST_NAME: Valerie
     STREET: 624 Randall
       CITY: Grove
      STATE: FL
        ZIP: 33321
 COMMISSION: 20542.50
       RATE: 0.05
*************************** 2. row ***************************
    REP_NUM: 35
  LAST_NAME: Hull
 FIRST_NAME: Richard
     STREET: 532 Jackson
       CITY: Sheldon
      STATE: FL
        ZIP: 33553
 COMMISSION: 39216.00
       RATE: 0.07
*************************** 3. row ***************************
    REP_NUM: 65
  LAST_NAME: Perez
 FIRST_NAME: Juan
     STREET: 1626 Taylor
       CITY: Fillmore
      STATE: FL
        ZIP: 33336
 COMMISSION: 23487.00
       RATE: 0.05
*************************** 4. row ***************************
    REP_NUM: 85
  LAST_NAME: Webb
 FIRST_NAME: Tina
     STREET: NULL
       CITY: NULL
      STATE: NULL
        ZIP: NULL
 COMMISSION: NULL
       RATE: NULL
4 rows in set (0.00 sec)
```

Option to display rows in the results vertically

Note: Going forward, the figures shown in this text will show query results in Oracle SQL*Plus. If you are using Oracle SQL*Plus Worksheet, Access, or MySQL, the data in your query results will be the same as the query results shown in the figures. However, your query results might be formatted differently.

■ Correcting Errors in the Database

After executing a SELECT command to view a table's data, you might find that you need to change the value in a column. You can use the **UPDATE** command shown in Figure 2.27 to update a value in a table. The first UPDATE command shown in the figure changes the last name in the row on which the sales rep number is 85 to Perry. The SELECT command displays the updated data.

FIGURE 2.27 Using an UPDATE command to change a value

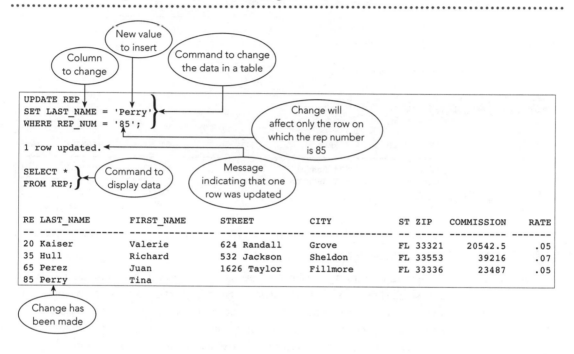

If you need to delete a record, you can use the **DELETE** command. The DELETE command shown in Figure 2.28 deletes any row on which the sales rep number is 85. The SELECT command displays the updated data.

FIGURE 2.28 Using a DELETE command to delete a row

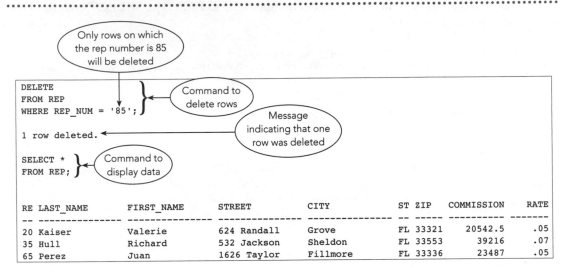

QUESTION   How do I correct errors in my data?

ANSWER   It depends on the type of error you need to correct. If you added a row that should not be in the database, use a DELETE command to remove it. If you forgot to add a row, you can use an INSERT command to add it. If you added a row that contained some incorrect data, you can use an UPDATE command to make the necessary corrections. Alternatively, you could use a DELETE command to remove the row containing the error and then use an INSERT command to insert the correct row.

■ Saving SQL Commands

Saving SQL commands in a file lets you use the commands again without retyping them. The exact manner in which you create and use saved files depends on the SQL implementation that you are using.

SQL*PLUS
USER

In Oracle SQL*Plus, you can use an editor to create a special file, called a **script file**, that contains one or more SQL commands. To create a script file in Oracle SQL*Plus, type EDIT followed by the name of the file you want to create. (Oracle assigns the file extension .sql automatically.) Type the command(s), save the file, and then close the editor. To run the command(s) in the file in Oracle SQL*Plus, type @ (the "at" symbol) followed by the filename. For example, to run a file named cre_cust, you would type @cre_cust. (If you saved the script file in a folder other than the default folder for your DBMS, you would need to include the full path to the file as part of the filename.) After you press the Enter key, Oracle executes the command(s) in the file.

SQL*PLUS
WORKSHEEET
USER

To create a script file in Oracle SQL*Plus Worksheet, type the command(s) in the upper pane of the program window, click File on the menu bar, click Save Input As, and then use the Save Worksheet As dialog box to specify a filename and the location in which to save the file. (Oracle assigns the file extension .sql automatically.) To run the command(s) in the file, click File on the menu bar, click Open, browse to and select the file, click the Open button, and then click the Execute button.

ACCESS
USER

In Access, create the query in SQL view and then click the Save button on the toolbar to save the query as an object in the database. To run the query without first viewing the SQL command, select the query in the Database window and then click the Open button. If you want to view the SQL command before running the query, select the query in the Database window and then click the Design button. The SQL command will be displayed; you can run the command by clicking the Run button on the toolbar.

MYSQL
USER

In MySQL, you can use an editor (such as Notepad) or a word processor to create text files containing your commands. (If you use a word processor, make sure to save the files as text files with .txt filename extensions.) To run the command(s) in the file from MySQL, type the word SOURCE or a slash and a period (\.) followed by the name of the file. For example, to run a file named cre_cust.txt, you would type one of the following commands:

```
SOURCE cre_cust.txt
```

or

```
\. cre_cust.txt
```

If you saved the script file in a folder other than the default folder for your DBMS, you would need to include the full path to the file as part of the filename. After you press the Enter key, MySQL executes the command(s) contained in the file.

■ Creating the Remaining Database Tables

To create the remaining tables in the Premiere Products database, you need to execute the appropriate CREATE TABLE and INSERT commands. You should save these commands on a hard drive or floppy disk so you can re-create your database, if necessary, by running the scripts (in Oracle or MySQL) or by opening the Access objects.

Figure 2.29 shows the CREATE TABLE command for the CUSTOMER table. Notice that the column CUSTOMER_NAME is specified as NOT NULL. Additionally, the CUSTOMER_NUM column is the table's primary key, indicating that the CUSTOMER_NUM column will be the unique identifier of rows within the table. With this column designated as the primary key, the DBMS will reject any attempt to store a customer number if that number already exists in the table.

FIGURE 2.29

CREATE TABLE command for the CUSTOMER table

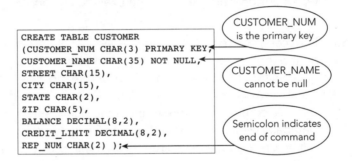

```
CREATE TABLE CUSTOMER
(CUSTOMER_NUM CHAR(3) PRIMARY KEY,
CUSTOMER_NAME CHAR(35) NOT NULL,
STREET CHAR(15),
CITY CHAR(15),
STATE CHAR(2),
ZIP CHAR(5),
BALANCE DECIMAL(8,2),
CREDIT_LIMIT DECIMAL(8,2),
REP_NUM CHAR(2) );
```

CUSTOMER_NUM is the primary key

CUSTOMER_NAME cannot be null

Semicolon indicates end of command

After creating the CUSTOMER table, you can create another file containing the INSERT commands to add the customer rows to the table. Each command must end with a semicolon. Figure 2.30 shows the INSERT commands to load the CUSTOMER table with data. As noted previously, to enter an apostrophe (single quotation mark) in the value for a field, type two single quotation marks, as illustrated in the name in the first INSERT command (Al's Appliance and Sport) in Figure 2.30.

FIGURE 2.30   INSERT commands for the CUSTOMER table

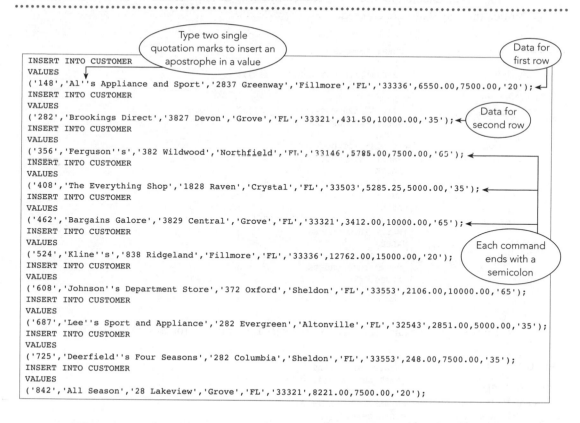

Figures 2.31 through 2.36 show files containing the CREATE TABLE and INSERT commands for creating and inserting data into the ORDERS, PART, and ORDER_LINE tables in the Premiere Products database. Figure 2.31 contains the CREATE TABLE command for the ORDERS table.

FIGURE 2.31

. . . .

CREATE TABLE
command for the
ORDERS table

```
CREATE TABLE ORDERS
(ORDER_NUM CHAR(5) PRIMARY KEY,
ORDER_DATE DATE,
CUSTOMER_NUM CHAR(3) );
```

ORDER_NUM
is the primary key

Figure 2.32 contains the INSERT commands to load data into the ORDERS table. Notice the way dates are entered.

FIGURE 2.32
. . . .
INSERT
commands for
the ORDERS
table

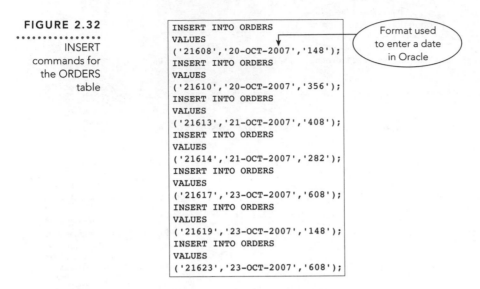

```
INSERT INTO ORDERS
VALUES
('21608','20-OCT-2007','148');
INSERT INTO ORDERS
VALUES
('21610','20-OCT-2007','356');
INSERT INTO ORDERS
VALUES
('21613','21-OCT-2007','408');
INSERT INTO ORDERS
VALUES
('21614','21-OCT-2007','282');
INSERT INTO ORDERS
VALUES
('21617','23-OCT-2007','608');
INSERT INTO ORDERS
VALUES
('21619','23-OCT-2007','148');
INSERT INTO ORDERS
VALUES
('21623','23-OCT-2007','608');
```

Format used to enter a date in Oracle

Figure 2.33 contains the CREATE TABLE command for the PART table.

FIGURE 2.33
. . . .
CREATE TABLE
command for the
PART table

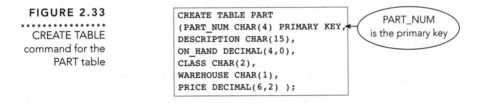

```
CREATE TABLE PART
(PART_NUM CHAR(4) PRIMARY KEY,
DESCRIPTION CHAR(15),
ON_HAND DECIMAL(4,0),
CLASS CHAR(2),
WAREHOUSE CHAR(1),
PRICE DECIMAL(6,2) );
```

PART_NUM is the primary key

Figure 2.34 contains the INSERT commands to load data into the PART table.

FIGURE 2.34

• • • • • • • • • • • • • •

INSERT
commands for
the PART table

```
INSERT INTO PART
VALUES
('AT94','Iron',50,'HW','3',24.95);
INSERT INTO PART
VALUES
('BV06','Home Gym',45,'SG','2',794.95);
INSERT INTO PART
VALUES
('CD52','Microwave Oven',32,'AP','1',165.00);
INSERT INTO PART
VALUES
('DL71','Cordless Drill',21,'HW','3',129.95);
INSERT INTO PART
VALUES
('DR93','Gas Range',8,'AP','2',495.00);
INSERT INTO PART
VALUES
('DW11','Washer',12,'AP','3',399.99);
INSERT INTO PART
VALUES
('FD21','Stand Mixer',22,'HW','3',159.95);
INSERT INTO PART
VALUES
('KL62','Dryer',12,'AP','1',349.95);
INSERT INTO PART
VALUES
('KT03','Dishwasher',8,'AP','3',595.00);
INSERT INTO PART
VALUES
('KV29','Treadmill',9,'SG','2',1390.00);
```

Figure 2.35 contains the CREATE TABLE command for the ORDER_LINE table. Notice the way the primary key is defined when it consists of more than one column.

FIGURE 2.35

• • • • • • • • • • • • • • •

CREATE TABLE
command for the
ORDER_LINE
table

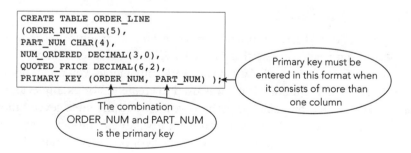

```
CREATE TABLE ORDER_LINE
(ORDER_NUM CHAR(5),
PART_NUM CHAR(4),
NUM_ORDERED DECIMAL(3,0),
QUOTED_PRICE DECIMAL(6,2),
PRIMARY KEY (ORDER_NUM, PART_NUM) );
```

Primary key must be entered in this format when it consists of more than one column

The combination ORDER_NUM and PART_NUM is the primary key

Figure 2.36 contains the INSERT commands to load data into the ORDER_LINE table.

FIGURE 2.36
....
INSERT com-
mands for the
ORDER_LINE
table

```
INSERT INTO ORDER_LINE
VALUES
('21608','AT94',11,21.95);
INSERT INTO ORDER_LINE
VALUES
('21610','DR93',1,495.00);
INSERT INTO ORDER_LINE
VALUES
('21610','DW11',1,399.99);
INSERT INTO ORDER_LINE
VALUES
('21613','KL62',4,329.95);
INSERT INTO ORDER_LINE
VALUES
('21614','KT03',2,595.00);
INSERT INTO ORDER_LINE
VALUES
('21617','BV06',2,794.95);
INSERT INTO ORDER_LINE
VALUES
('21617','CD52',4,150.00);
INSERT INTO ORDER_LINE
VALUES
('21619','DR93',1,495.00);
INSERT INTO ORDER_LINE
VALUES
('21623','KV29',2,1290.00);
```

■ Describing a Table

The CREATE TABLE command defines a table's structure by indicating its columns, data types, and column lengths. The CREATE TABLE command also indicates which columns cannot accept nulls. When you work with a table, you might not have access to the CREATE TABLE command that was used to create it. For example, another programmer might have created the table, or perhaps you created the table several months ago but did not save the command. You still might want to examine the table's structure to see details concerning the columns in the table. Each DBMS provides a method to examine a table's structure.

ORACLE
USER

In either Oracle SQL*Plus or Oracle SQL*Plus Worksheet, you can use the **DESCRIBE** command to list all the columns in a table and their corresponding data types. Figure 2.37 shows the table structure for the REP table. Notice that the command doesn't end with a semicolon because DESCRIBE is not a standard SQL command.

FIGURE 2.37   DESCRIBE command for the REP table (Oracle)

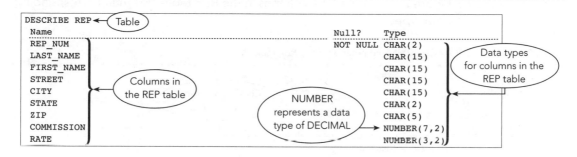

In Figure 2.37, Oracle uses the NUMBER data type instead of the DECIMAL data type specified in the CREATE TABLE command.

ACCESS USER : In Access, you use the **Documenter** tool to produce documentation about tables and other objects stored in a database. To start the Documenter, click Tools on the menu bar, point to Analyze, and then click Documenter. In the Documenter dialog box, select the tables (and other objects if desired) for which you want to produce documentation, and then click the OK button. The Object Definition window opens and displays a report containing the requested documentation. Figure 2.38 shows the printed version of the documentation for the REP table. You can customize the Documenter to control the amount of detail included in the report. The report shown in Figure 2.38 contains minimal detail.

FIGURE 2.38

Documentation for the REP table (Access)

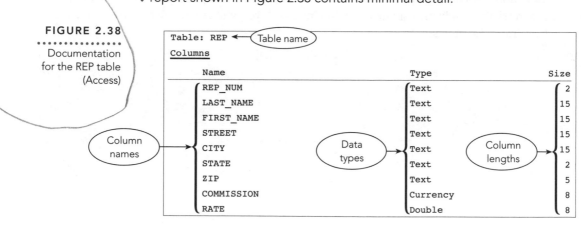

In MySQL, you can use the SHOW COLUMNS command shown in Figure 2.39 to list all the columns in the REP table and their corresponding data types.

FIGURE 2.39 SHOW COLUMNS command for the REP table (MySQL)

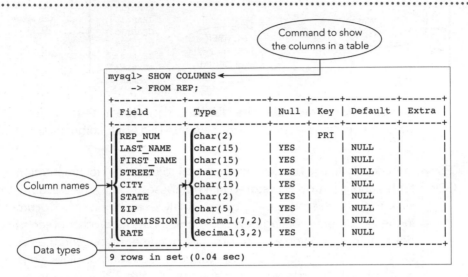

In this chapter, you learned how to create and run SQL commands, create tables, and add rows to these tables. In the next two chapters, you will use SQL commands to create and run queries that retrieve data from these tables. In Chapter 3, you will create queries that retrieve data from a single table. In Chapter 4, you will create queries that retrieve data from multiple tables.

■ SUMMARY

- A relation is a two-dimensional table in which the entries are single-valued (each location in the table contains a single entry), each column has a distinct name (or attribute name), all values in a column are values of the same attribute, the order of the rows and columns is immaterial, and each row contains unique values.

- A relational database is a collection of relations. Columns in a table (relation) are also called fields or attributes. Rows are also called records or tuples.

- An entity is a person, place, object, event, or idea for which you want to store and process data. An attribute is a characteristic or property of an entity. A relationship is the association between entities.

- Use the CREATE TABLE command to create a table by typing the table name and then listing within a single set of parentheses the columns in the table.

- The primary key of a table is the column or collection of columns that uniquely identifies a given row in the table. When writing a table's structure, you usually identify primary keys by underlining the column name(s).

- Use the DROP TABLE command to delete a table and all its data from the database.

- The common data types in SQL are INTEGER, SMALLINT, DECIMAL, CHAR, and DATE.

- A null data value (or null) is a special value that is used when the actual value for a column is unknown, unavailable, or not applicable.

- Use the NOT NULL clause in a CREATE TABLE command to identify columns that cannot accept null values.

- Use the INSERT command to load data into a table.

- Use the SELECT command to view the data in a table.

- Use the UPDATE command to change the value in a column.

- Use the DELETE command to delete a row in a table.

- You can use the DESCRIBE command (Oracle SQL*Plus or Oracle SQL*Plus Worksheet), the Documenter tool (Access), or the SHOW COLUMNS command (MySQL) to display a table's structure and layout.

attribute
buffer
column
command buffer
CREATE TABLE
data type
DELETE
DESCRIBE
Documenter
DROP TABLE
entity
field
INSERT
Microsoft Access
MySQL
NOT NULL
null
null data value

one-to-many relationship
Oracle SQL*Plus
Oracle SQL*Plus Worksheet
primary key
qualify
record
relation
relational database
relationship
repeating group
script file
SELECT
SQL
statement history
Structured Query Language
tuple
UPDATE

■ REVIEW QUESTIONS

1. What is a relation?

2. What is a relational database?

3. What are other terms for rows in a relation? What are other terms for columns in a relation?

4. How do you create a table in SQL?

5. What is a primary key? How do you identify it in the shorthand representation of a relation?

6. How do you delete a table?

7. What are the common data types in SQL?

8. What is a null value? How do you identify columns that cannot accept null values?

9. Which command do you use to add a row to a table?

10. Which command do you use to view the data in a table?

11. Which command do you use to change the value in a column in a table?

12. Which command do you use to delete rows from a table?

13. How do you display or print the layout of a table in Oracle SQL*Plus, Oracle SQL*Plus Worksheet, Access, or MySQL?

ORACLE
USER
If you are using SQL*Plus or SQL*Plus Worksheet for these exercises and want to print your commands and results, first save the commands and results in a file. To do so, execute a command consisting of the word SPOOL followed by the name of the file you want to create. All the commands (and their results) that you enter from that point will be saved in the file that you named. For example, to save the commands and results to a file named CHAPTER2.SQL on drive A, execute the following command before beginning your work:

```
SPOOL A:CHAPTER2.SQL
```

When you have finished entering and running your commands, execute the SPOOL OFF command. Then start any program that can open text files, open the file that you saved, and print it using the Print command on the File menu.

ACCESS
USER
If you are using Access for these exercises and want to print your commands and results, open Word and create a new document into which you will copy the commands and results. To place a command in the Word document, use the mouse to select the entire command, click the Copy button on the toolbar in Access, activate your Word document, and then click the Paste button on the toolbar in Word. To place results that appear in a datasheet into the Word document, select the entire datasheet by right-clicking its selector (the box in the upper left corner of the datasheet), clicking Copy on the shortcut menu, activating your Word document, and then clicking the Paste button on the Word toolbar to paste the datasheet into the document.

MYSQL
USER
If you are using MySQL for these exercises and wish to print a copy of your commands and results, first save the commands and results in a file. To do so, execute a command consisting of the word TEE or a slash and the letter T (\T) followed by the name of the file. All the commands (and their results) that you enter from that point will be saved in the file that you named. For example, to save the commands and results to a file named CHAPTER2.TXT on drive A, execute either of the following commands before beginning your work:

```
TEE A:CHAPTER2.TXT
```

or

```
\T A:CHAPTER2.TXT
```

When you have finished entering and running your commands, start any program that can open text files, open the file that you saved, and print it using the Print command on the File menu.

■EXERCISES (Premiere Products)

Use SQL and Figure 2.1 to complete the following exercises.

1. Execute the CREATE TABLE commands to create the tables in the Premiere Products database. (The CREATE TABLE commands you need are shown in Figures 2.3, 2.29, 2.31, 2.33, and 2.35.)

2. Add the sales reps shown in Figure 2.1 to the REP table using INSERT commands.

3. Add the customers shown in Figure 2.1 to the CUSTOMER table.

4. Add the orders shown in Figure 2.1 to the ORDERS table.

5. Add the part data shown in Figure 2.1 to the PART table.

6. Add the order line information shown in Figure 2.1 to the ORDER_LINE table.

■EXERCISES (Henry Books)

Use SQL and Figures 1.4 through 1.7 in Chapter 1 to complete the following exercises.

1. Use the appropriate CREATE TABLE commands to create the tables in the Henry Books database. The information you need appears in Figure 2.40.

2. Add the branch information shown in Figure 1.4 (in Chapter 1) to the BRANCH table using INSERT commands.

3. Add the publisher information shown in Figure 1.4 to the PUBLISHER table.

4. Add the author information shown in Figure 1.5 to the AUTHOR table.

5. Add the book information shown in Figure 1.6 to the BOOK table. In the PAPERBACK column, enter Y where the entry in Figure 1.6 is Yes and N where the entry is No.

6. Add the author and book information shown in Figure 1.7 to the WROTE table.

7. Add the inventory information shown in Figure 1.7 to the INVENTORY table.

FIGURE 2.40   Table layouts for the Henry Books database

BRANCH

Column	Type	Length	Decimal Places	Nulls Allowed?	Description
BRANCH_NUM	Decimal	2	0	No	Branch number (primary key)
BRANCH_NAME	Char	50			Branch name
BRANCH_LOCATION	Char	50			Branch location
NUM_EMPLOYEES	Decimal	2	0		Number of employees

PUBLISHER

Column	Type	Length	Decimal Places	Nulls Allowed?	Description
PUBLISHER_CODE	Char	3		No	Publisher code (primary key)
PUBLISHER_NAME	Char	25			Publisher name
CITY	Char	20			Publisher city

AUTHOR

Column	Type	Length	Decimal Places	Nulls Allowed?	Description
AUTHOR_NUM	Decimal	2	0	No	Author number (primary key)
AUTHOR_LAST	Char	12			Author last name
AUTHOR_FIRST	Char	10			Author first name

BOOK

Column	Type	Length	Decimal Places	Nulls Allowed?	Description
BOOK_CODE	Char	4		No	Book code (primary key)
TITLE	Char	40			Book title
PUBLISHER_CODE	Char	3			Publisher code
TYPE	Char	3			Book type
PRICE	Decimal	4	2		Book price
PAPERBACK	Char	1			Paperback (Y, N)

Column	Type	Length	Decimal Places	Nulls Allowed?	Description
BOOK_CODE	Char	4		No	Book code (primary key)
AUTHOR_NUM	Decimal	2	0	No	Author number (primary key)
SEQUENCE	Decimal	1	0		Sequence number

INVENTORY

Column	Type	Length	Decimal Places	Nulls Allowed?	Description
BOOK_CODE	Char	4		No	Book code (primary key)
BRANCH_NUM	Decimal	2	0	No	Branch number (primary key)
ON_HAND	Decimal	2	0		Units on hand

■EXERCISES (Alexamara Marina Group)

Use SQL and Figures 1.8 through 1.12 in Chapter 1 to complete the following exercises.

1. Use the appropriate CREATE TABLE commands to create the tables in the Alexamara Marina Group database. The information you need appears in Figure 2.41 on the following pages.

2. Add the marina information in Figure 1.8 to the MARINA table using INSERT commands.

3. Add the owner information in Figure 1.9 to the OWNER table using INSERT commands.

4. Add the slip information in Figure 1.10 to the MARINA_SLIP table using INSERT commands.

5. Add the service category information in Figure 1.11 to the SERVICE_CATEGORY table using INSERT commands.

6. Add the service request information in Figure 1.12 to the SERVICE_REQUEST table using INSERT commands. *Hint:* The values for next service dates on some rows are null. To enter a null value, you can use the type of INSERT command illustrated in Example 4. Alternatively, you might find it simpler to enter a date that does not occur on any other row (such as December 31, 2010) for the dates that should be null. Once you have added all rows, you can change this date to null on any row on which it occurs. To do so in Oracle, the command would be:

```
UPDATE SERVICE_REQUEST
SET NEXT_SERVICE_DATE = Null
WHERE NEXT_SERVICE_DATE =
    '31-DEC-2010';
```

To do so in Access, the command would be:

```
UPDATE SERVICE_REQUEST
SET NEXT_SERVICE_DATE = Null
WHERE NEXT_SERVICE_DATE =
    #12/31/2010#;
```

To do so in MySQL, the command would be:

```
UPDATE SERVICE_REQUEST
SET NEXT_SERVICE_DATE = Null
WHERE NEXT_SERVICE_DATE =
    '2010-12-31';)
```

FIGURE 2.41   Table layouts for the Alexamara Marina Group database

MARINA

Column	Type	Length	Decimal Places	Nulls Allowed?	Description
MARINA_NUM	Char	4		No	Marina number (primary key)
NAME	Char	20			Marina name
ADDRESS	Char	15			Marina street address
CITY	Char	15			Marina city
STATE	Char	2			Marina state
ZIP	Char	5			Marina zip code

OWNER

Column	Type	Length	Decimal Places	Nulls Allowed?	Description
OWNER_NUM	Char	4		No	Owner number (primary key)
LAST_NAME	Char	50			Owner last name
FIRST_NAME	Char	20			Owner first name
ADDRESS	Char	15			Owner street address
CITY	Char	15			Owner city
STATE	Char	2			Owner state
ZIP	Char	5			Owner zip code

MARINA_SLIP

Column	Type	Length	Decimal Places	Nulls Allowed?	Description
SLIP_ID	Decimal	4	0	No	Slip ID (primary key)
MARINA_NUM	Char	4			Marina number
SLIP_NUM	Char	4			Slip number in the marina
LENGTH	Decimal	4	0		Length of slip (in feet)
RENTAL_FEE	Decimal	8	2		Annual rental fee for the slip
BOAT_NAME	Char	50			Name of boat currently in the slip
BOAT_TYPE	Char	50			Type of boat currently in the slip
OWNER_NUM	Char	4			Number of boat owner renting the slip

SERVICE_CATEGORY

Column	Type	Length	Decimal Places	Nulls Allowed?	Description
CATEGORY_NUM	Decimal	4	0	No	Category number (primary key)
CATEGORY_DESCRIPTION	Char	255			Category description

SERVICE_REQUEST

Column	Type	Length	Decimal Places	Nulls Allowed?	Description
SERVICE_ID	Decimal	4	0	No	Service ID (primary key)
SLIP_ID	Decimal	4	0		Slip ID of the boat for which service is requested
CATEGORY_NUM	Decimal	4	0		Category number of the requested service
DESCRIPTION	Char	255			Description of specific service required for boat
STATUS	Char	255			Description of status of service request
EST_HOURS	Decimal	4	2		Estimated number of hours required to complete the service
SPENT_HOURS	Decimal	4	2		Hours already spent on the service
NEXT_SERVICE_DATE	Date				Next scheduled date for work on this service (or null if no next service date is specified)

72

CHAPTER 3

Single-Table Queries

OBJECTIVES

- Retrieve data from a database using SQL commands
- Use compound conditions
- Use computed columns
- Use the SQL LIKE operator
- Use the SQL IN operator
- Sort data using the ORDER BY clause
- Sort data using multiple keys and in ascending and descending order
- Use SQL aggregate functions
- Use subqueries
- Group data using the GROUP BY clause
- Select individual groups of data using the HAVING clause
- Retrieve columns with null values

Introduction

In this chapter, you will learn about the SQL SELECT command that is used to retrieve data in a database. You will examine ways to sort data and use SQL functions to count rows and calculate totals. You also will learn about a special feature of SQL that lets you nest SELECT commands by placing one SELECT command inside another. Finally, you will learn how to group rows that have matching values in some column.

ORACLE USER : You might need to change the number of characters displayed on the screen to keep your data from wrapping to the next line. In Oracle SQL*Plus and Oracle SQL*Plus Worksheet, you can change the number of characters displayed on each line using the SET LINESIZE command. To change the line size to 100 (100 characters per line), you would execute the following command:

```
SET LINESIZE 100
```

You might also need to change the number of lines that are displayed on the screen. To change the page size to 50 (50 lines per screen), you would execute the following command:

```
SET PAGESIZE 50
```

You would execute these commands prior to executing your SELECT command. Once you have executed the SET LINESIZE and/or the SET PAGESIZE commands, the changes will remain in effect throughout your entire session.

ACCESS USER : In Access, the query results appear in Datasheet view (see Figure 2.24 in Chapter 2). If the data displayed does not fit on the screen, you can adjust the columns to best fit the data they contain by double-clicking the right edge of each column heading. You can use the scroll bars to view data that has scrolled off the screen.

MYSQL USER : By default, MySQL displays query results by arranging rows in a horizontal presentation (see Figure 2.25 in Chapter 2). If the horizontal presentation causes the data to wrap to the next line, you can choose to display the data vertically rather than horizontally. To do so, replace the semicolon at the end of the command with a slash (\) and the letter G (see Figure 2.26 in Chapter 2).

■ Constructing Simple Queries

One of the most important features of a database management system is its ability to answer a wide variety of questions concerning the data in a database. When you need to find data that answers a specific question, you use a query. A **query** is simply a question represented in a way that the DBMS can understand.

In SQL, you use the SELECT command to query a database. The basic form of the SQL SELECT command is SELECT-FROM-WHERE. After you type the word SELECT, you list the columns that you want to include in the query results. This portion of the command is called the **SELECT clause**. Next, you type the word FROM followed by the name of the table that contains the data you need to query. This portion of the command is called the **FROM clause**. Finally, after the word WHERE, you list any conditions (restrictions) that apply to the data you want to retrieve. This optional portion of the command is called the **WHERE clause**. For example, if you need to retrieve the rows for only those customers with credit limits of $7,500, include a condition in the WHERE clause specifying that the value in the CREDIT_LIMIT column must be $7500 (CREDIT_LIMIT = 7500).

There are no special formatting rules in SQL. In this text, the FROM clause and the WHERE clause (when it is used) appear on separate lines only to make the commands more readable and understandable.

Note: The various implementations of SQL might display query output differently. For example, the format of column headings and numeric data might differ. The SQL implementation used in this text truncates (shortens) the column headings to fit the width of the column data. Although your output should contain the same data that appears in the figures in this text, its format might differ slightly.

Retrieving Certain Columns and All Rows

You can write a command to retrieve specified columns and all rows from a table, as shown in Example 1.

EXAMPLE 1   List the number, name, and balance of all customers.

Because you need to list *all* customers, you won't need to include a WHERE clause; you don't need to put any restrictions on the data to retrieve. You simply list the columns to be included (CUSTOMER_NUM, CUSTOMER_NAME, and BALANCE) in the SELECT clause and the name of the table (CUSTOMER) in the FROM clause. The query and its results appear in Figure 3.1.

FIGURE 3.1
· · · · · · · · · · · · · · · ·
SELECT com-
mand to select
certain columns

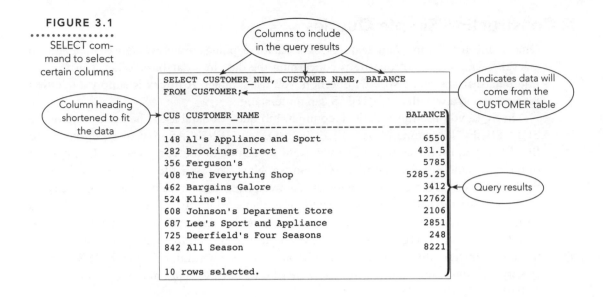

Columns to include
in the query results

Indicates data will
come from the
CUSTOMER table

Column heading
shortened to fit
the data

Query results

```
SELECT CUSTOMER_NUM, CUSTOMER_NAME, BALANCE
FROM CUSTOMER;

CUS CUSTOMER_NAME                          BALANCE
--- --------------------------------    ----------
148 Al's Appliance and Sport                 6550
282 Brookings Direct                        431.5
356 Ferguson's                               5785
408 The Everything Shop                   5285.25
462 Bargains Galore                          3412
524 Kline's                                 12762
608 Johnson's Department Store              2106
687 Lee's Sport and Appliance              2851
725 Deerfield's Four Seasons                 248
842 All Season                              8221

10 rows selected.
```

Retrieving All Columns and All Rows

You can use the same type of command as shown in Example 1 to retrieve all columns and all rows from a table. However, as Example 2 illustrates, you can use a shortcut to accomplish this task.

EXAMPLE 2   ⋮   List the complete PART table.

Instead of listing every column in the SELECT clause, you can use an asterisk (*) to indicate that you want to include all columns. The result will list all columns in the order in which you described them to the system when you created the table. If you want the columns listed in a different order, type the column names in the order in which you want them to appear in the query results. In this case, assuming that the default order is appropriate, you can use the query shown in Figure 3.2.

FIGURE 3.2
· · · · · · · · · · · · · · · · ·
SELECT com-
mand to select all
columns

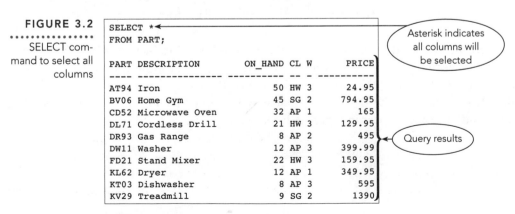

```
SELECT *
FROM PART;

PART DESCRIPTION        ON_HAND CL W      PRICE
---- ---------------    ------- -- - ----------
AT94 Iron                   50 HW 3      24.95
BV06 Home Gym               45 SG 2     794.95
CD52 Microwave Oven         32 AP 1        165
DL71 Cordless Drill         21 HW 3     129.95
DR93 Gas Range               8 AP 2        495
DW11 Washer                 12 AP 3     399.99
FD21 Stand Mixer            22 HW 3     159.95
KL62 Dryer                  12 AP 1     349.95
KT03 Dishwasher              8 AP 3        595
KV29 Treadmill               9 SG 2       1390
```

Using a WHERE Clause

You use the WHERE clause to retrieve rows that satisfy some condition, as shown in Example 3.

 EXAMPLE 3 : What is the name of customer number 148?

You can use the WHERE clause to restrict the query output to customer number 148, as shown in Figure 3.3. Because CUSTOMER_NUM is a character column, the value 148 is enclosed in single quotation marks. In addition, because the CUSTOMER_NUM column is the primary key of the CUSTOMER table, there can be only one customer whose number matches the number in the WHERE clause.

FIGURE 3.3

· · · · · · · · · · · · · · · ·

SELECT command to find the name of customer number 148

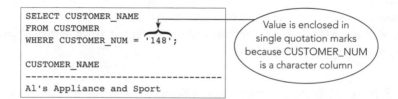

```
SELECT CUSTOMER_NAME
FROM CUSTOMER
WHERE CUSTOMER_NUM = '148';

CUSTOMER_NAME
------------------------------------
Al's Appliance and Sport
```

Value is enclosed in single quotation marks because CUSTOMER_NUM is a character column

The condition in the preceding WHERE clause is called a simple condition. A **simple condition** has this form: column name, comparison operator, and then either another column name or a value. Figure 3.4 lists the comparison operators that you can use in SQL. Notice that there are two versions of the "not equal to" operator: < > and !=. You must use the correct one for your SQL implementation. (If you use the wrong operator, your system will generate an error, in which case, you'll know to use the other version.)

FIGURE 3.4

· · · · · · · · · · · · · · · ·

Comparison operators used in SQL commands

Comparison Operator	Description
=	Equal to
<	Less than
>	Greater than
<=	Less than or equal to
>=	Greater than or equal to
< >	Not equal to (used by most SQL implementations)
!=	Not equal to (used by some SQL implementations)

EXAMPLE 4 : Find the number and name of each customer located in the city of Grove.

The only difference between this example and the previous one is that in Example 3, there could not be more than one row in the answer because the condition involved the table's primary key. In Example 4, the condition involves a column that is not the table's primary key. Because there is more than one customer located in the city of Grove, the results can and do contain more than one row, as shown in Figure 3.5.

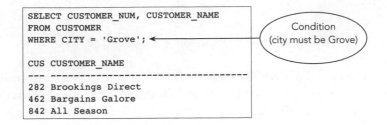

```
SELECT CUSTOMER_NUM, CUSTOMER_NAME
FROM CUSTOMER
WHERE CITY = 'Grove';

CUS CUSTOMER_NAME
--- ----------------------------------
282 Brookings Direct
462 Bargains Galore
842 All Season
```

Condition
(city must be Grove)

Note: In general, SQL is not case-sensitive; you can type commands using uppercase or lowercase letters. There is one exception to this rule, however. When you are dealing with character values, you must use the correct case. For example, if you use WHERE CITY = 'grove', some SQL implementations will not select any rows because "Grove" is considered to be different from "grove."

EXAMPLE 5 : Find the number, name, balance, and credit limit for all customers with credit limits that exceed their balances.

A simple condition also can involve a comparison of two columns, as shown in Figure 3.6. The WHERE clause uses a comparison operator to select those rows in which the balance is greater than the credit limit.

FIGURE 3.6   SELECT command to find all customers with balances that exceed their credit limits
....

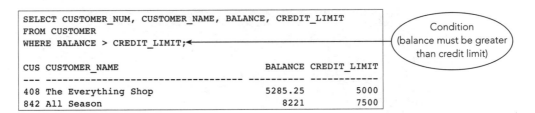

```
SELECT CUSTOMER_NUM, CUSTOMER_NAME, BALANCE, CREDIT_LIMIT
FROM CUSTOMER
WHERE BALANCE > CREDIT_LIMIT;

CUS CUSTOMER_NAME                          BALANCE CREDIT_LIMIT
--- ----------------------------------    ------- ------------
408 The Everything Shop                   5285.25         5000
842 All Season                               8221         7500
```

Condition
(balance must be greater
than credit limit)

Using Compound Conditions

As stated previously, the conditions you have seen so far are called simple conditions. The following examples require compound conditions. You form **compound conditions** by connecting two or more simple conditions using the AND, OR, and NOT operators. When the AND operator connects simple conditions, all the simple conditions must be true in order for the compound condition to be true. When the OR operator connects simple conditions, the compound condition will be true whenever any one of the simple conditions is true. Preceding a condition by the NOT operator reverses the truth of the original condition. For example, if the original condition is true, the new condition will be false; if the original condition is false, the new one will be true.

EXAMPLE 6   List the descriptions of all parts that are located in warehouse 3 and for which there are more than 25 units on hand.

In this example, you need to retrieve those parts that meet *both* conditions—the warehouse number is equal to 3 *and* the number of units on hand is greater than 25. To find the answer, you form a compound condition using the AND operator, as shown in Figure 3.7. The query examines the data in the PART table and lists the parts that are located in warehouse 3 and for which there are more than 25 units on hand.

FIGURE 3.7
....
SELECT
command with
an AND
condition

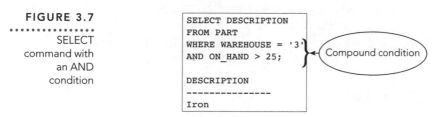

For readability, each of the simple conditions in the query shown in Figure 3.7 appears on a separate line. Some people prefer to put the conditions on the same line with parentheses around each simple condition, as shown in Figure 3.8. These two methods accomplish the same thing. In this text, simple conditions will appear on separate lines and without parentheses.

FIGURE 3.8
....
SELECT
command with
WHERE clause
and an AND
condition on a
single line

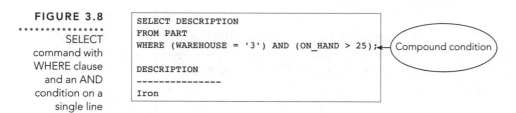

EXAMPLE 7   List the descriptions of all parts that are located in warehouse 3 or for which there are more than 25 units on hand.

In this example, you need to retrieve those parts for which the warehouse number is equal to 3, *or* the number of units on hand is greater than 25, *or* both. To do this, you form a compound condition using the OR operator, as shown in Figure 3.9.

FIGURE 3.9
· · · · · · · · · · · · · · ·
SELECT
command with
an OR condition

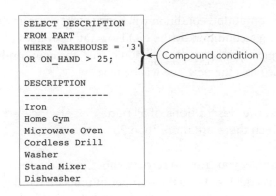

```
SELECT DESCRIPTION
FROM PART
WHERE WAREHOUSE = '3'
OR ON_HAND > 25;

DESCRIPTION
----------------
Iron
Home Gym
Microwave Oven
Cordless Drill
Washer
Stand Mixer
Dishwasher
```

Compound condition

EXAMPLE 8 : List the descriptions of all parts that are not in warehouse 3.

For this example, you could use a simple condition and the "not equal to" operator (WHERE WAREHOUSE < > '3'). As an alternative, you could use the EQUAL operator (=) in the condition and precede the entire condition with the NOT operator, as shown in Figure 3.10.

FIGURE 3.10
· · · · · · · · · · · · · · ·
SELECT
command with a
NOT condition

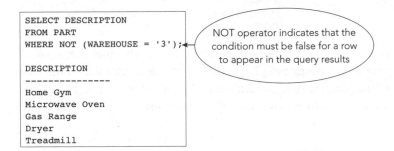

```
SELECT DESCRIPTION
FROM PART
WHERE NOT (WAREHOUSE = '3');

DESCRIPTION
----------------
Home Gym
Microwave Oven
Gas Range
Dryer
Treadmill
```

NOT operator indicates that the condition must be false for a row to appear in the query results

You don't need to enclose the condition WAREHOUSE = '3' in parentheses, but doing so makes the command more readable. Notice that by phrasing the condition in this form, you avoid the problem of determining whether your SQL implementation uses the < > or != not equal to operator.

Using the BETWEEN Operator

Example 9 requires a compound condition to determine the answer.

EXAMPLE 9 : List the number, name, and balance of all customers with balances greater than or equal to $2,000 and less than or equal to $5,000.

You can use a WHERE clause and the AND operator as shown in Figure 3.11 to retrieve the data.

FIGURE 3.11
• • • • • • • • • • • • • •
SELECT
command with
an AND
condition for a
single column

```
SELECT CUSTOMER_NUM, CUSTOMER_NAME, BALANCE
FROM CUSTOMER
WHERE BALANCE >= 2000
AND BALANCE <= 5000;

CUS CUSTOMER_NAME                               BALANCE
--- --------------------------------- ----------
462 Bargains Galore                              3412
608 Johnson's Department Store                   2106
687 Lee's Sport and Appliance                    2851
```

An alternative to this approach uses the BETWEEN operator, as shown in Figure 3.12.

FIGURE 3.12   SELECT command with the BETWEEN operator
• •

```
SELECT CUSTOMER_NUM, CUSTOMER_NAME, BALANCE
FROM CUSTOMER
WHERE BALANCE BETWEEN 2000 AND 5000;

CUS CUSTOMER_NAME                               BALANCE
--- --------------------------------- ----------
462 Bargains Galore                              3412
608 Johnson's Department Store                   2106
687 Lee's Sport and Appliance                    2851
```

BETWEEN operator
indicates that the value
must be between the
listed numbers

The BETWEEN operator is not an essential feature of SQL; you have just seen that you can obtain the same result without it. Using the BETWEEN operator, however, does make certain SELECT commands simpler to construct.

Note: The BETWEEN operator is inclusive, meaning that a value equal to either end would be selected. In the clause BETWEEN 2000 and 5000, for example, values of either 2,000 or 5,000 would make the condition true.

Using Computed Columns

You can use computed columns in SQL queries. A **computed column** is a column that does not exist in the database but can be computed using data in the existing columns. Computations can involve any arithmetic operator shown in Figure 3.13.

FIGURE 3.13
• • • • • • • • • • • • • • •
Arithmetic
operators

Arithmetic Operator	Description
+	Addition
–	Subtraction
*	Multiplication
/	Division

EXAMPLE 10 Find the number, name, and available credit (the credit limit minus the balance) for each customer.

There is no column for available credit in the Premiere Products database, but you can compute the available credit from two columns that are present: CREDIT_LIMIT and BALANCE. To compute the available credit, you use the expression CREDIT_LIMIT – BALANCE, as shown in Figure 3.14.

FIGURE 3.14 SELECT command with a computed column

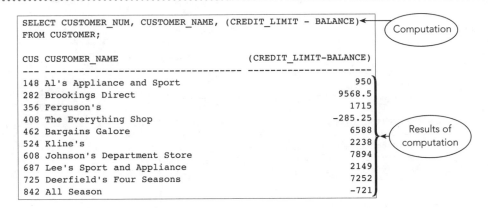

The parentheses around the calculation (CREDIT_LIMIT – BALANCE) are not essential but improve readability.

Some SQL implementations use special headings for computed columns, such as SUM1 or COUNT2. You can also assign a name to a computed column by following the computation with the word AS and the desired name. The command in Figure 3.15, for example, assigns the name AVAILABLE_CREDIT to the computation.

FIGURE 3.15 SELECT command with a named computation

```
SELECT CUSTOMER_NUM, CUSTOMER_NAME, (CREDIT_LIMIT - BALANCE) AS AVAILABLE_CREDIT
FROM CUSTOMER;

CUS CUSTOMER_NAME                          AVAILABLE_CREDIT
--- -------------------------------------- ----------------
148 Al's Appliance and Sport                            950
282 Brookings Direct                                 9568.5
356 Ferguson's                                          1715
408 The Everything Shop                              -285.25
462 Bargains Galore                                     6588
524 Kline's                                             2238
608 Johnson's Department Store                          7894
687 Lee's Sport and Appliance                           2149
725 Deerfield's Four Seasons                            7252
842 All Season                                          -721
```

Name of computed column

Note: You can use names containing spaces following the word AS. In many SQL implementations, including Oracle SQL*Plus, Oracle SQL*Plus Worksheet, and MySQL, you do so by enclosing the name in quotation marks (for example, AS "AVAILABLE CREDIT"). Other SQL implementations require you to enclose the name in other special characters. For example, in Access you would enclose the name in square brackets (AS [AVAILABLE CREDIT]).

EXAMPLE 11 : Find the number, name, and available credit for each customer with at least $5,000 of available credit.

You can use computed columns in comparisons, as shown in Figure 3.16.

FIGURE 3.16   SELECT command with a computation in the condition

```
SELECT CUSTOMER_NUM, CUSTOMER_NAME, (CREDIT_LIMIT - BALANCE) AS AVAILABLE_CREDIT
FROM CUSTOMER
WHERE (CREDIT_LIMIT - BALANCE) >= 5000;          Computation
                                                 in condition

CUS CUSTOMER_NAME                    AVAILABLE_CREDIT
--- -------------------------------- ----------------
282 Brookings Direct                          9568.5
462 Bargains Galore                             6588
608 Johnson's Department Store                  7894
725 Deerfield's Four Seasons                    7252
```

Using the LIKE Operator

In most cases, your conditions will involve exact matches, such as retrieving rows for each customer located in the city of Grove. In some cases, however, exact matches will not work. For example, you might know that the desired value contains only a certain collection of characters. In such cases, you use the LIKE operator with a wildcard symbol, as shown in Example 12.

EXAMPLE 12 : List the number, name, and complete address of each customer located on a street that contains the letters "Central".

All you know is that the addresses you want contain a certain collection of characters ("Central") somewhere in the STREET column, but you don't know where. In SQL, the percent sign (%) is used as a wildcard to represent any collection of characters. As shown in Figure 3.17, the condition LIKE '%Central%' will retrieve information for each customer whose street contains some collection of characters, followed by the letters "Central," followed potentially by some additional characters. Note that this query also would retrieve information for a customer whose address is "123 Centralia" because "Centralia" also contains the letters "Central".

FIGURE 3.17   SELECT command with a LIKE operator and wildcards

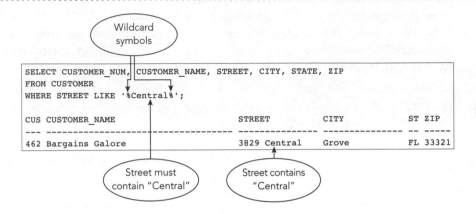

Another wildcard symbol is the underscore (_), which represents any individual character. For example, "T_m" represents the letter "T" followed by any single character, followed by the letter "m," and would retrieve rows that include the words Tim, Tom, or T3m, for example.

Note: In a large database, you should use wildcards only when absolutely necessary. Searches involving wildcards can be extremely slow to process.

Aᴄᴄᴇss **:** Access uses different wildcard symbols. The symbol for any collection of
USER **:** characters is the asterisk (*), as shown in Figure 3.18. The symbol for an
: individual character is the question mark (?).

FIGURE 3.18   Access SELECT command with wildcards

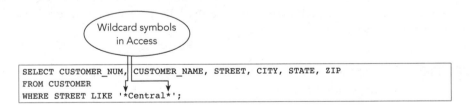

Using the IN Operator

An **IN clause**, which consists of the IN operator followed by a collection of values, provides a concise way of phrasing certain conditions, as Example 13 illustrates. You will see another use for the IN clause in more complex examples later in this chapter.

EXAMPLE 13 : List the number, name, and credit limit for each customer with a credit
limit of $5,000, $10,000, or $15,000.

In this query, you can use an IN clause to determine whether a credit limit is $5,000, $10,000, or $15,000. You could obtain the same answer by using the condition WHERE CREDIT_LIMIT = 5000 OR CREDIT_LIMIT = 10000 OR CREDIT_LIMIT = 15000. The approach shown in Figure 3.19 is simpler because the IN clause contains a collection of values: 5000, 10000, and 15000. The condition is true for those rows in which the value in the CREDIT_LIMIT column is in this collection.

FIGURE 3.19 SELECT command with an IN clause

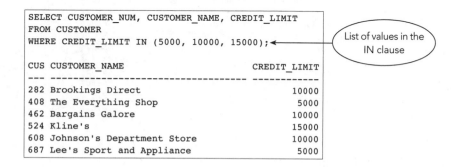

■ Sorting

Recall that the order of rows in a table is immaterial to the DBMS. From a practical stand-point, this means that when you query a relational database, there is no defined order in which the results are displayed. Rows can be displayed in the order in which the data was originally entered, but even this is not certain. If the order in which the data is displayed is important, you can *specifically* request that the results appear in a desired order. In SQL, you specify the results order by using the **ORDER BY clause**.

Using the ORDER BY Clause

You use the ORDER BY clause to list data in a specific order, as shown in Example 14.

EXAMPLE 14 : List the number, name, and balance of each customer. Order (sort) the
output in ascending (increasing) order by balance.

The column on which data is to be sorted is called a **sort key** or simply a **key**. In this case, you need to order the output by balance, so the sort key is the BALANCE column. To sort the output, use an ORDER BY clause, followed by the sort key. If you don't specify a sort order, the default is ascending. The query appears in Figure 3.20.

FIGURE 3.20 SELECT command to sort rows

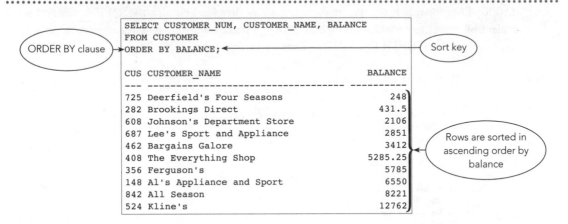

Additional Sorting Options

Sometimes you might need to sort data by more than one key, as shown in Example 15.

EXAMPLE 15 : List the number, name, and credit limit of each customer. Order the customers by name within descending credit limit. (In other words, sort the customers by credit limit in descending order. Within each group of customers that have a common credit limit, sort the customers by name in ascending order.)

This example involves two new ideas: sorting on multiple keys—CREDIT_LIMIT and CUSTOMER_NAME—and using descending order for one of the keys. When you need to sort data on two columns, the more important column (in this case, CREDIT_LIMIT) is called the **major sort key** (or the **primary sort key**) and the less important column (in this case, CUSTOMER_NAME) is called the **minor sort key** (or the **secondary sort key**). To sort on multiple keys, you list the keys in order of importance in the ORDER BY clause. To sort in descending order, you follow the name of the sort key with the word DESC, as shown in Figure 3.21.

FIGURE 3.21   SELECT command to sort data using multiple sort keys

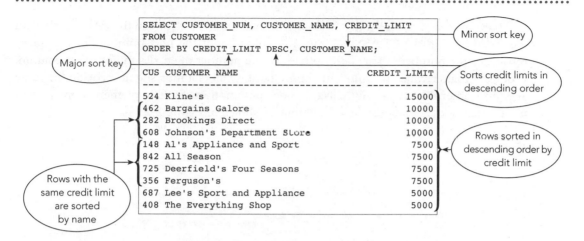

Using Functions

SQL has special functions, called **aggregate functions**, to calculate sums, averages, counts, maximum values, and minimum values. These functions apply to groups of rows. They could apply to all the rows in a table (for example, calculating the average balance of all customers). They could also apply to those rows satisfying some particular condition (for example, the average balance of all customers of sales rep 20). The descriptions of the aggregate functions appear in Figure 3.22.

FIGURE 3.22

SQL aggregate functions

Function	Description
AVG	Calculates the average value in a column
COUNT	Determines the number of rows in a table
MAX	Determines the maximum value in a column
MIN	Determines the minimum value in a column
SUM	Calculates a total of the values in a column

Using the COUNT Function

The COUNT function, as illustrated in Example 16, counts the number of rows in a table.

EXAMPLE 16 : How many parts are in item class HW?

For this query, you need to determine the total number of rows in the PART table with the value HW in the CLASS column. You could count the part numbers in the query results, or the number of part descriptions, or the number of entries in any other column. It doesn't matter which column you choose because all columns should provide the same answer. Rather than arbitrarily selecting one column, most SQL implementations let you use the asterisk (*) to represent any column, as shown in Figure 3.23.

FIGURE 3.23
• • • • • • • • • • • • • • • •
SELECT
command to
count rows

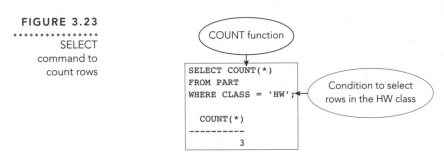

If your SQL implementation does not permit the use of the asterisk, you can write the query as follows:

```
SELECT COUNT(PART_NUM)
FROM PART
WHERE CLASS = 'HW';
```

Using the SUM Function

If you want to calculate the total of all customers' balances, you can use the SUM function, as illustrated in Example 17.

EXAMPLE 17 : Find the total number of Premiere Products customers and the total of
their balances.

When you use the SUM function, you must specify the column to total, and the column data type must be numeric. (How could you calculate a sum of names or addresses?) The query appears in Figure 3.24.

FIGURE 3.24
• • • • • • • • • • • • • • • •
SELECT
command to
count records
and calculate
a total

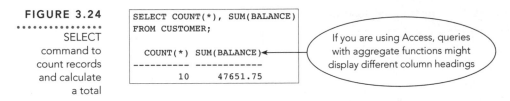

Using the AVG, MAX, and MIN functions is similar to using SUM, except that different statistics are calculated. Figure 3.25 shows a query that illustrates these functions.

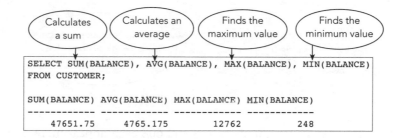

Note: When you use the SUM, AVG, MAX, or MIN functions, SQL ignores any null value(s) in the column and eliminates them from the computations.

Note: Null values in numeric columns can produce strange results when statistics are computed. Suppose that the BALANCE column accepts null values, that there are currently four customers with rows in the CUSTOMER table, and that their respective balances are $100, $200, $300, and null (unknown). When you calculate the average balance, most SQL implementations will ignore the null value and obtain $200 (($100 + $200 + $300) / 3). Similarly, if you calculate the total of the balances, SQL ignores the null value and calculates a total of $600. If you count the number of customers in the table, however, SQL includes the row containing the null value, and the result is 4. Thus the total of the balances ($600) divided by the number of customers (4) would give $150 as the average balance!

Using the DISTINCT Operator

In some situations, the DISTINCT operator is useful when used in conjunction with the COUNT function. Before examining such a situation, you need to understand how to use the DISTINCT operator. Examples 18 and 19 illustrate the most common uses of DISTINCT.

EXAMPLE 18 : Find the number of each customer that currently has an open order (that is, an order currently in the ORDERS table).

The command seems fairly simple. If a customer currently has an open order, there must be at least one row in the ORDERS table on which that customer's number appears. You could use the query shown in Figure 3.26 to find the customer numbers with open orders.

FIGURE 3.26
· · · · · · · · · · · · · · · ·
Customer
numbers with
open orders

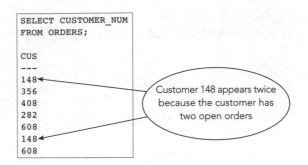

Notice that customer numbers 148 and 608 each appear more than once in the output; they both currently have more than one open order in the ORDERS table. Suppose you want to list each customer only once, as illustrated in Example 19.

EXAMPLE 19 : Find the number of each customer that currently has an open order. List each customer only once.

To ensure uniqueness, you can use the DISTINCT operator, as shown in Figure 3.27.

FIGURE 3.27
· · · · · · · · · · · · · · · ·
Results without
repeated
customer
numbers

```
SELECT DISTINCT(CUSTOMER_NUM)
FROM ORDERS;

CUS
---
148
282
356
408
608
```

You might be wondering about the relationship between COUNT and DISTINCT, because both involve counting rows. The following example identifies the differences.

EXAMPLE 20 : Count the number of customers that currently have open orders.

The query shown in Figure 3.28 counts the number of customers using the CUSTOMER_NUM column.

FIGURE 3.28
· · · · · · · · · · · · · · · ·
Count that
includes
repeated
customer
numbers

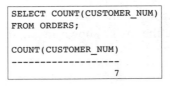

QUESTION   What's wrong with the query results shown in Figure 3.28?

ANSWER   The answer, 7, is the result of counting the customers that have open orders multiple times—once for each separate order currently on file. The result counts each customer number and does not eliminate redundant customer numbers to provide an accurate count of the number of customers.

Some SQL implementations, including Oracle and MySQL (but not Access), allow you to use the DISTINCT operator to calculate the correct count, as shown in Figure 3.29.

FIGURE 3.29
••••••••••••••••
Count without repeated customer numbers (using DISTINCT within COUNT)

```
SELECT COUNT(DISTINCT(CUSTOMER_NUM))
FROM ORDERS;

COUNT(DISTINCT(CUSTOMER_NUM))
----------------------------
                           5
```

■ Nesting Queries

Sometimes obtaining the results you need requires two or more steps, as shown in the next two examples.

EXAMPLE 21 ⋮ List the number of each part in class AP.

The command to obtain the answer is shown in Figure 3.30.

FIGURE 3.30
••••••••••••••••
Selecting all parts in class AP

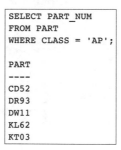

```
SELECT PART_NUM
FROM PART
WHERE CLASS = 'AP';

PART
----
CD52
DR93
DW11
KL62
KT03
```

EXAMPLE 22 ⋮ List the order numbers that contain an order line for a part in class AP.

You need to find those order numbers in the ORDER_LINE table that correspond to the part numbers in the results of the previous query. After viewing those results (CD52, DR93, DW11, KL62, and KT03), you can use the command shown in Figure 3.31.

FIGURE 3.31

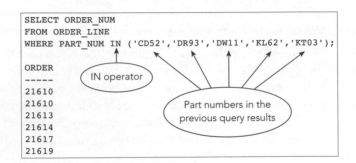

Query using the
previous query's
results

Subqueries

It is possible to place one query inside another. The inner query is called a **subquery**; the subquery is evaluated first. After the subquery has been evaluated, the outer query can use the results of the subquery to find its results, as shown in Example 23.

EXAMPLE 23 : Find the answer to Examples 21 and 22 in one step.

You can find the same result as in the previous two examples in a single step by using a subquery. In Figure 3.32, the command shown in parentheses is the subquery. This subquery is evaluated first, producing a temporary table. The temporary table is used only to evaluate the query—it is not available to the user or displayed—and is deleted after the evaluation of the query is complete. In this example, the temporary table has only a single column (PART_NUM) and five rows (CD52, DR93, DW11, KL62, and KT03). The outer query is evaluated next. In this case, the outer query will retrieve the order number on every row in the ORDER_LINE table for which the part number is in the results of the subquery. Because that table contains only the part numbers in class AP, you obtain the desired list of order numbers.

FIGURE 3.32
Using the IN
operator and a
subquery

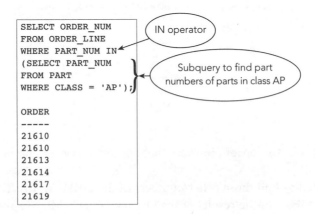

MYSQL   The command shown in Figure 3.32 will not work in versions of MySQL
     USER   prior to 4.1. In older versions, you would need to accomplish this task in
      two steps as illustrated in Examples 21 and 22.

EXAMPLE 24   List the number, name, and balance for each customer whose balance
       exceeds the average balance.

In this case, you use a subquery to obtain the average balance. Because this subquery produces a single number, you can compare each customer's balance with this number, as shown in Figure 3.33.

FIGURE 3.33

Query using an
operator and a
subquery

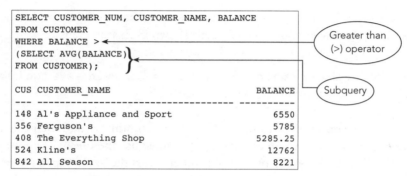

```
SELECT CUSTOMER_NUM, CUSTOMER_NAME, BALANCE
FROM CUSTOMER
WHERE BALANCE >
(SELECT AVG(BALANCE)
FROM CUSTOMER);


CUS CUSTOMER_NAME                        BALANCE
--- ------------------------------------ ----------
148 Al's Appliance and Sport                6550
356 Ferguson's                              5785
408 The Everything Shop                   5285.25
524 Kline's                                12762
842 All Season                              8221
```

Greater than
(>) operator

Subquery

Note: You cannot use the condition BALANCE > AVG(BALANCE) in the WHERE clause; you must use a subquery to obtain the average balance. Then you can use the results of the subquery in a condition, as illustrated in Figure 3.33.

MYSQL   The command shown in Figure 3.33 will not work in versions of MySQL
     USER   prior to 4.1. In older versions, you would need to accomplish this task in
      two steps. First create and run a query to calculate the average balance.
      You will use this calculation in the second query. Then create and run a
      second query to select those customers whose balance is greater than
      the value you calculated in the first query.

■ Grouping

Grouping creates groups of rows that share some common characteristic. If customers are grouped by credit limit, for example, the first group would contain customers with $5,000 credit limits, the second group would contain customers with $7,500 credit limits, and so on. If, on the other hand, customers are grouped by sales rep number, the first group would contain those customers represented by sales rep number 20, the second group would contain those customers represented by sales rep number 35, and the third group would contain those customers represented by sales rep number 65.

When you group rows, any calculations indicated in the SELECT command are performed for the entire group. For example, if customers are grouped by rep number and the query requests the average balance, the results include the average balance for the group of customers represented by rep number 20, the average balance for the group represented by rep number 35, and the average balance for the group represented by rep number 65. The following examples illustrate this process.

Using the GROUP BY Clause

The **GROUP BY clause** lets you group data on a particular column, such as REP_NUM, and then calculate statistics, if desired, as shown in Example 25.

EXAMPLE 25 : For each sales rep, list the rep number and the average balance of the rep's customers.

Because you need to group customers by rep number and then calculate the average balance for all customers in each group, you must use the GROUP BY clause. In this case, GROUP BY REP_NUM puts all customers with the same rep number into a separate group. Any statistics requested in the SELECT command are calculated for each group. It is important to note that the GROUP BY clause does not sort the data in a particular order; you must use the ORDER BY clause to sort data. Assuming that the report should be ordered by rep number, you can use the command shown in Figure 3.34.

FIGURE 3.34
••••••••••••••••
Grouping records
using a column

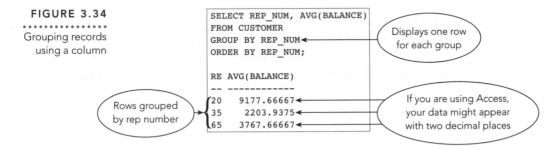

When rows are grouped, one line of output is produced for each group. The only things that can be displayed are statistics calculated for the group or columns whose values are the same for all rows in a group.

QUESTION   Is it appropriate to display the rep number?

ANSWER   Yes, because the rep number in one row in a group must be the same as
            the rep number in any other row in the group.

QUESTION   Would it be appropriate to display a customer number?

ANSWER   No, because the customer number varies from one row in a group to
            another. (The same rep is associated with many customers.) SQL would not
            be able to determine which customer number to display for the group. The
            system displays an error message if you attempt to list a customer number.

■ Using a HAVING Clause

The **HAVING clause** is used to restrict the groups that will be included, as shown in
Example 26.

EXAMPLE 26 : Repeat the previous example, but list only those reps who represent fewer
              than four customers.

    The only difference between Examples 25 and 26 is the restriction to display only
those reps who represent fewer than four customers. This restriction does not apply to
individual rows but rather to *groups*. Because the WHERE clause applies only to rows,
you cannot use it to accomplish the kind of selection that is required. Fortunately, the
HAVING clause does for groups what the WHERE clause does for rows. In Figure 3.35,
the row created for a group will be displayed only if the count of the number of rows in the
group sum is less than 4; in addition, all groups will be ordered by rep number.

FIGURE 3.35
Restricting the
groups to include

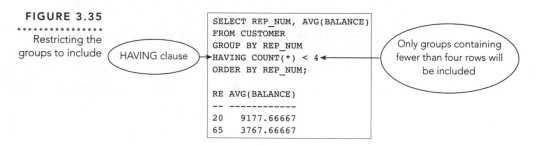

```
SELECT REP_NUM, AVG(BALANCE)
FROM CUSTOMER
GROUP BY REP_NUM
HAVING COUNT(*) < 4
ORDER BY REP_NUM;

RE AVG(BALANCE)
-- ------------
20    9177.66667
65    3767.66667
```

HAVING clause

Only groups containing
fewer than four rows will
be included

HAVING vs. WHERE

Just as you can use the WHERE clause to limit the *rows* that are included in a query's
result, you can use the HAVING clause to limit the *groups* that are included. The following
examples illustrate the difference between these two clauses.

EXAMPLE 27 : List each credit limit and the number of customers having each credit limit.

To count the number of customers that have a given credit limit, you must group the data by credit limit, as shown in Figure 3.36.

FIGURE 3.36
....
Counting the
number of rows
in each group

```
SELECT CREDIT_LIMIT, COUNT(*)
FROM CUSTOMER
GROUP BY CREDIT_LIMIT;

CREDIT_LIMIT    COUNT(*)
------------ ----------
        5000          2
        7500          4
       10000          3
       15000          1
```

EXAMPLE 28 : Repeat Example 27, but list only those credit limits held by more than one customer.

Because this condition involves a group total, the query includes a HAVING clause, as shown in Figure 3.37.

FIGURE 3.37
....
Displaying
groups that
contain more
than one row

```
SELECT CREDIT_LIMIT, COUNT(*)
FROM CUSTOMER
GROUP BY CREDIT_LIMIT
HAVING COUNT(*) > 1;

CREDIT_LIMIT    COUNT(*)
------------ ----------
        5000          2
        7500          4
       10000          3
```

EXAMPLE 29 : List each credit limit and the total number of customers of sales rep 20 that have this limit.

The condition involves only rows, so using the WHERE clause is appropriate, as shown in Figure 3.38.

FIGURE 3.38
....
Restricting the
rows to be
grouped

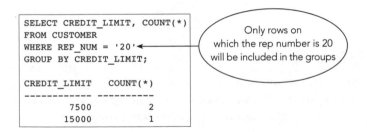

```
SELECT CREDIT_LIMIT, COUNT(*)
FROM CUSTOMER
WHERE REP_NUM = '20'
GROUP BY CREDIT_LIMIT;

CREDIT_LIMIT    COUNT(*)
------------ ----------
        7500          2
       15000          1
```

Only rows on
which the rep number is 20
will be included in the groups

EXAMPLE 30 : Repeat Example 29, but list only those credit limits held by more than one customer.

Because the conditions involve rows and groups, you must use both a WHERE clause and a HAVING clause, as shown in Figure 3.39.

FIGURE 3.39
....
Restricting the
rows and the
groups

```
SELECT CREDIT_LIMIT, COUNT(*)
FROM CUSTOMER
WHERE REP_NUM = '20'
GROUP BY CREDIT_LIMIT
HAVING COUNT(*) > 1;

CREDIT_LIMIT    COUNT(*)
------------ ----------
        7500           2
```

Only rows on which the rep number is 20 will be included in the groups

Only groups containing more than one row will be included in the results

In Example 30, rows from the original table are considered only if the sales rep number is 20. These rows are then grouped by credit limit and the count is calculated. Only groups for which the calculated count is greater than 1 are displayed.

■ Nulls

Sometimes a condition involves a column that can be null, as illustrated in Example 31.

EXAMPLE 31 : List the number and name of each customer with a null (unknown) street value.

You might expect the condition to be something like STREET = NULL. The correct format is actually STREET IS NULL, as shown in Figure 3.40. (To select a customer whose street is not null, you use the condition STREET IS NOT NULL.) In the current Premiere Products database, no customer has a null street value; therefore no rows are retrieved in the query results.

FIGURE 3.40
....
Selecting rows
containing null
values in the
STREET column

```
SELECT CUSTOMER_NUM, CUSTOMER_NAME
FROM CUSTOMER
WHERE STREET IS NULL;

no rows selected
```

In this chapter, you learned how to create queries that retrieve data from a single table by constructing appropriate SELECT commands. In the next chapter, you will learn how to create queries that retrieve data from multiple tables. The queries you created in this chapter used the clauses and operators shown in Figure 3.41.

FIGURE 3.41   SQL query clauses and operators

Clause or Operator	Description
AND operator	Specifies that all simple conditions must be true for the compound condition to be true
BETWEEN operator	Specifies a range of values in a condition
DESC operator	Sorts the query results in descending order based on the column name
DISTINCT operator	Ensures uniqueness in the condition by eliminating redundant values
FROM clause	Indicates the table from which to retrieve the specified columns
GROUP BY clause	Groups rows based on the specified column
HAVING clause	Limits a condition to the groups that are included
IN operator	Finds a value in a group of values specified in the condition
IS NOT NULL operator	Finds rows that do not contain a null value in the specified column
IS NULL operator	Finds rows that contain a null value in the specified column
LIKE operator	Indicates a pattern of characters to find in a condition
NOT operator	Reverses the truth or falsity of the original condition
OR operator	Specifies that the compound condition is true whenever any of the simple conditions is true
ORDER BY clause	Lists the query results in the specified order based on the column name
SELECT clause	Specifies the columns to retrieve in the query
WHERE clause	Specifies any conditions for the query

■ SUMMARY

- The basic form of the SQL SELECT command is SELECT-FROM-WHERE. Specify the columns to be listed after the word SELECT (or type * to select all columns), and then specify the table name that contains these columns after the word FROM. Optionally, you can include conditions after the word WHERE.

- Simple conditions are written in the following form: column name, comparison operator, column name or value. Simple conditions can involve any of the comparison operators: =, >, >=, <, <=, or < > or != (not equal to).

- You can form compound conditions by combining simple conditions using the operators AND, OR, and NOT.

- Use the BETWEEN operator to indicate a range of values in a condition.

- Use computed columns in SQL commands by using arithmetic operators and writing the computation in place of a column name. You can assign a name to the computation by following the computation with the word AS and then the desired name.

- To check for a value in a character column that is similar to a particular string of characters, use the LIKE operator. The percent (%) wildcard represents any collection of characters. The underscore (_) wildcard represents any single character. In Access, the asterisk (*) wildcard represents any collection of characters and the question mark (?) wildcard represents any single character.

- To determine whether a column contains one of a particular set of values, use the IN operator.

- Use an ORDER BY clause to sort data. List sort keys in order of importance. To sort in descending order, follow the sort key with the word DESC.

- SQL processes the aggregate functions COUNT, SUM, AVG, MAX, and MIN. These calculations apply to groups of rows.

- To avoid duplicates in a query that uses an aggregate function, precede the column name with the DISTINCT operator.

- When one SQL query is placed inside another, it is called a subquery. The inner query (the subquery) is evaluated first.

- Use a GROUP BY clause to group data.

- Use a HAVING clause to restrict the output to certain groups.

- Use the IS NULL operator in a WHERE clause to find rows containing a null value in some column. Use the IS NOT NULL operator in a WHERE clause to find rows that do not contain a null value.

■ KEY TERMS

aggregate function
compound condition
computed column
FROM clause
GROUP BY clause
grouping
HAVING clause
IN clause
key
major sort key

minor sort key
ORDER BY clause
primary sort key
query
secondary sort key
SELECT clause
simple condition
sort key
subquery
WHERE clause

■ REVIEW QUESTIONS

1. Describe the basic form of the SQL SELECT command.

2. What is the form of a simple condition?

3. How do you form a compound condition?

4. In SQL, what operator do you use to determine whether a value is between two other values without using an AND condition?

5. How do you use a computed column in SQL? How do you name the computation?

6. In which clause would you use a wild-card in a condition?

7. What wildcards are available in SQL and what do they represent?

8. How do you determine whether a column contains one of a particular set of values without using AND in the condition?

9. How do you sort data?

10. How do you sort data on more than one sort key? What is the more important key called? What is the less important key called?

11. How do you sort data in descending order?

12. What are the SQL aggregate functions?

13. How do you avoid duplicates in the results of SQL queries?

14. What is a subquery?

15. How do you group data in SQL queries?

16. When grouping, how can you restrict the output to only those groups satisfying some condition?

17. How do you find rows in which a partic-ular column contains a null value?

■ EXERCISES (Premiere Products)

Use SQL and the Premiere Products database (see Figure 1.2 in Chapter 1) to complete the following exercises. Use the Notes at the end of Chapter 2 to print your output if directed to do so by your instructor.

1. List the part number, description, and price for all parts.

2. List all rows and columns for the complete ORDERS table.

3. List the names of customers with credit limits of $7,500 or less.

4. List the order number for each order placed by customer number 148 on 10/20/2007. (*Hint*: If you need help, use the discussion of the DATE data type in Figure 2.13 in Chapter 2.)

5. List the number and name of each cus-tomer represented by sales rep 35 or sales rep 65.

6. List the part number and part description of each part that is not in item class SG.

7. List the part number, description, and number of units on hand for each part that has between 10 and 25 units on hand, including both 10 and 25. Do this two ways.

8. List the part number, part description, and on-hand value (units on hand * unit price) of each part in item class AP. (On-hand value is really units on hand * cost, but there is no COST column in the PART table.) Assign the name ON_HAND_VALUE to the computation.

9. List the part number, part description, and on-hand value for each part whose on-hand value is at least $7,500. Assign the name ON_HAND_VALUE to the computation.

10. Use the IN operator to list the part number and part description of each part in item class AP or SG.

11. Find the number and name of each customer whose name begins with the letter "K."

12. List all details about all parts. Order the output by part description.

13. List all details about all parts. Order the output by part number within item class. (That is, order the output by item class and then by part number.)

14. How many customers have balances that are more than their credit limits?

15. Find the total of the balances for all customers represented by sales rep 65 that have balances that are less than their credit limits.

16. List the part number, part description, and on-hand value of each part whose number of units on hand is more than the average number of units on hand for all parts. (*Hint:* Use a subquery.)

17. What is the price of the most expensive part in the database?

18. What is the part number, description, and price of the most expensive part in the database? (*Hint:* Use a subquery.)

19. List the sum of the balances of all customers for each sales rep. Order and group the results by sales rep number.

20. List the sum of the balances of all customers for each sales rep, but restrict the output to those sales reps for whom the sum is more than $10,000.

21. List the part number of any part with an unknown description.

■EXERCISES (Henry Books)

Use SQL and the Henry Books database (Figures 1.4 through 1.7 in Chapter 1) to complete the following exercises. Use the Notes at the end of Chapter 2 to print your output if directed to do so by your instructor.

1. List the book code and book title of each book.

2. List the complete PUBLISHER table.

3. List the name of each publisher located in Boston.

4. List the name of each publisher not located in Boston.

5. List the name of each branch that has at least nine employees.

6. List the book code and book title of each book that has the type SFI.

7. List the book code and book title of each book that has the type SFI and is in paperback.

8. List the book code and book title of each book that has the type SFI or is published by the publisher with the publisher code SC.

9. List the book code, book title, and price of each book with a price between $20 and $30.

10. List the book code and book title of each book that has the type MYS and a price of less than $20.

11. Customers who are part of a special program get a 10% discount off regular book prices. List the book code, book title, and discounted price of each book. Use DISCOUNTED_PRICE as the name for the computed column, which should calculate 90% of the current price; that is, 100% less a 10% discount.

12. Find the name of each publisher containing the word "and." (*Hint:* Be sure that your query selects only those publishers that contain the word "and" and not those that contain the letters "and" in the middle of a word. For example, your query should select the publisher named "Farrar Straus and Giroux," but should *not* select the publisher named "Random House.")

13. List the book code and book title of each book that has the type SFI, MYS, or ART. Use the IN operator in your command.

14. Repeat Exercise 13, but also list the books in alphabetical order by title.

15. Repeat Exercise 13, but also include the price and list the books in descending order by price. Within a group of books having the same price, further order the books by title.

16. Display the list of book types in the database. List each book type only once.

17. How many books have the type SFI?

18. For each type of book, list the type and the average price.

19. Repeat Exercise 18, but consider only paperback books.

20. Repeat Exercise 18, but consider only paperback books for those types for which the average price is more than $10.

21. What is the most expensive book in the database?

22. What are the title(s) and price(s) of the least expensive book(s) in the database?

23. How many employees does Henry Books have?

■ EXERCISES (Alexamara Marina Group)

Use SQL and the Alexamara Marina Group database (Figures 1.8 through 1.12 in Chapter 1) to complete the following exercises. Use the Notes at the end of Chapter 2 to print your output if directed to do so by your instructor.

1. List the owner number, last name, and first name of every boat owner.

2. List the complete MARINA table (all rows and all columns).

3. List the last name and first name of every owner located in Bowton.

4. List the last name and first name of every owner not located in Bowton.

5. List the marina number and slip number for every slip whose length is equal to or less than 30 feet.

6. List the marina number and slip number for every boat with the type Dolphin 28.

7. List the slip number for every boat with the type Dolphin 28 that is located in marina 1.

8. List the boat name for each boat located in a slip whose length is between 25 and 30 feet.

9. List the slip number for every slip in marina 1 whose rental fee is less than $3,000.00.

10. Labor is billed at the rate of $60.00 per hour. List the slip ID, category number, estimated hours, and estimated labor cost for every service request. To obtain the estimated labor cost, multiply the estimated hours by 60. Use the column name ESTIMATED_COST for the estimated labor cost.

11. List the marina number and slip number for all slips containing a boat with the type Sprite 4000, Sprite 3000, or Ray 4025.

12. List the marina number, slip number, and boat name for all boats. Sort the results by boat name within marina number.

13. How many Dolphin 28 boats are stored at both marinas?

14. Calculate the total rental fees Alexamara receives each year based on the length of the slip.

CHAPTER 4

Multiple-Table Queries

OBJECTIVES

- Use joins to retrieve data from more than one table

- Use the IN and EXISTS operators to query multiple tables

- Use a subquery within a subquery

- Use an alias

- Join a table to itself

- Perform set operations (union, intersection, and difference)

- Use the ALL and ANY operators in a query

- Perform special operations (inner join, outer join, and product)

Introduction

In this chapter, you will learn how to retrieve data from two or more tables using one SQL command. You will join tables together and examine how similar results are obtained using the IN and EXISTS operators. Then you will use aliases to simplify queries and join a table to itself. You will also implement the set operations of union, intersection, and difference using SQL commands. You will examine two related SQL operators: ALL and ANY. Finally, you will perform inner joins, outer joins, and products.

■ Querying Multiple Tables

In Chapter 3, you learned how to retrieve data from a single table. Many queries require you to retrieve data from two or more tables, which requires that you join them. Then you formulate a query using the same commands that you use for single-table queries.

Note: In the following queries, the order of your results may be different from the ones in the text, although they should have the same rows. If order is important, you can include an ORDER BY clause in the query to ensure the results have the desired order.

Joining Two Tables

To retrieve data from more than one table, you must **join** the tables together by finding rows in the two tables that have identical values in matching columns. You can join tables by using a condition in the WHERE clause, as you will see in Example 1.

EXAMPLE 1 : List the number and name of each customer, together with the number, last name, and first name of the sales rep who represents the customer.

Because the customers' numbers and names are in the CUSTOMER table and the sales reps' numbers and names are in the REP table, you need to include both tables in the same SQL command so you can retrieve data from both tables. To join (relate) the tables, you'll construct the SQL command as follows:

1. In the SELECT clause, list all columns you want to display.

2. In the FROM clause, list all tables involved in the query.

3. In the WHERE clause, list the condition that restricts the data to be retrieved to only those rows from the two tables that match; that is, restrict it to the rows that have common values in matching columns.

As you learned in Chapter 2, it is often necessary to qualify a column to specify the particular column you are referencing. Qualifying column names is especially important when joining tables because you must join tables on *matching* columns that frequently have identical column names. To qualify a column name, precede the name of the column with the name of the table, followed by a period. The matching columns in this example are both named REP_NUM: there is a column in the REP table named REP_NUM and a column in the CUSTOMER table that also is named REP_NUM. The REP_NUM column in the REP table is written as REP.REP_NUM and the REP_NUM column in the CUSTOMER table is written as CUSTOMER.REP_NUM. The query and its results appear in Figure 4.1.

FIGURE 4.1
Joining tables

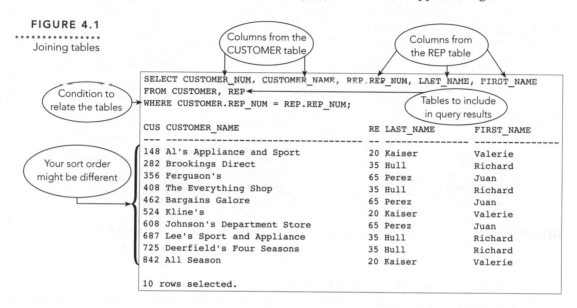

QUESTION   In the first row of output in Figure 4.1, the customer number is 148 and the customer name is Al's Appliance and Sport. These values represent the first row of the CUSTOMER table. Why is the sales rep number 20, the last name of the sales rep Kaiser, and the first name Valerie?

ANSWER   In the CUSTOMER table, the sales rep number for customer number 148 is 20. (This indicates that customer number 148 is *related* to sales rep number 20.) In the REP table, the last name of sales rep number 20 is Kaiser and the first name is Valerie.

When there is potential ambiguity in listing column names, you *must* qualify the columns involved in the query. It is permissible to qualify other columns as well, even if there is no possible confusion. Some people prefer to qualify all column names; in this text, however, you will qualify column names only when it is necessary.

EXAMPLE 2   : List the number and name of each customer whose credit limit is $7,500, together with the number, last name, and first name of the sales rep who represents the customer.

In Example 1, you used a condition in the WHERE clause only to relate a customer with a sales rep to join the tables. Although relating a customer with a sales rep is essential in this example as well, you also need to restrict the output to only those customers whose credit limits are $7,500. You can restrict the rows by using a compound condition, as shown in Figure 4.2.

FIGURE 4.2   Restricting the rows in a join

```
SELECT CUSTOMER_NUM, CUSTOMER_NAME, REP.REP_NUM, LAST_NAME, FIRST_NAME
FROM CUSTOMER, REP
WHERE CUSTOMER.REP_NUM = REP.REP_NUM          ← Condition to relate
AND CREDIT_LIMIT = 7500;                          the tables


CUS CUSTOMER_NAME                     RE LAST_NAME        FIRST_NAME
--- ------------------------------   -- ---------------   ---------------
148 Al's Appliance and Sport         20 Kaiser            Valerie
356 Ferguson's                       65 Perez             Juan        ← Condition to
725 Deerfield's Four Seasons         35 Hull              Richard       restrict the rows
842 All Season                       20 Kaiser            Valerie
```

EXAMPLE 3   For every part on order, list the order number, part number, part description, number of units ordered, quoted price, and unit price.

A part is considered "on order" if there is a row in the ORDER_LINE table in which the part appears. You can find the order number, number of units ordered, and quoted price in the ORDER_LINE table. To find the part description and the unit price, however, you need to look in the PART table. Then you need to find rows in the ORDER_LINE table and rows in the PART table that match (rows containing the same part number). The query and its results appear in Figure 4.3.

FIGURE 4.3   Joining the ORDER_LINE and PART tables

```
SELECT ORDER_NUM, ORDER_LINE.PART_NUM, DESCRIPTION, NUM_ORDERED, QUOTED_PRICE, PRICE
FROM ORDER_LINE, PART
WHERE ORDER_LINE.PART_NUM = PART.PART_NUM;


ORDER PART DESCRIPTION      NUM_ORDERED QUOTED_PRICE    PRICE
----- ---- ---------------  ----------- ------------  ----------
21608 AT94 Iron                      11        21.95      24.95
21610 DR93 Gas Range                  1          495        495
21610 DW11 Washer                     1       399.99     399.99
21613 KL62 Dryer                      4       329.95     349.95
21614 KT03 Dishwasher                 2          595        595
21617 BV06 Home Gym                   2       794.95     794.95
21617 CD52 Microwave Oven             4          150        165
21619 DR93 Gas Range                  1          495        495
21623 KV29 Treadmill                  2         1290       1390
```

QUESTION   Can you use PART.PART_NUM instead of ORDER_LINE.PART_NUM in the SELECT clause?

ANSWER   Yes. The values for these two columns match because they must satisfy the condition ORDER_LINE.PART_NUM = PART.PART_NUM.

Comparing JOIN, IN, and EXISTS

You join tables in SQL by including a condition in the WHERE clause to ensure that matching columns contain equal values (for example, ORDER_LINE.PART_NUM = PART.PART_NUM). You can obtain similar results by using either the IN operator (used in Chapter 3) or the EXISTS operator with a subquery. The choice is a matter of personal preference; either approach obtains the same results. The following examples illustrate the use of each operator.

EXAMPLE 4 ⋮ Find the description of each part included in order number 21610.

Because this query also involves retrieving data from the ORDER_LINE and PART tables as illustrated in Example 3, you could approach it in a similar fashion. There are two basic differences, however, between Examples 3 and 4. First, the query in Example 4 does not require as many columns; second, it involves only order number 21610. Having fewer columns to retrieve means that there will be fewer columns listed in the SELECT clause. You can restrict the query to a single order by adding the condition ORDER_NUM = '21610' to the WHERE clause. The query and its results appear in Figure 4.4.

FIGURE 4.4

Restricting rows when joining the ORDER_LINE and PART tables

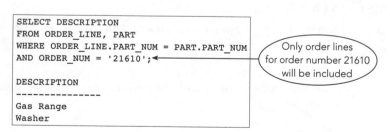

```
SELECT DESCRIPTION
FROM ORDER_LINE, PART
WHERE ORDER_LINE.PART_NUM = PART.PART_NUM
AND ORDER_NUM = '21610';

DESCRIPTION
---------------
Gas Range
Washer
```

Only order lines for order number 21610 will be included

Notice that the ORDER_LINE table is listed in the FROM clause, even though you don't need to display any columns from the ORDER_LINE table. The WHERE clause contains columns from the ORDER_LINE table, so it is necessary to include the table in the FROM clause.

Using IN

Another way to retrieve data from multiple tables in a query is to use the IN operator with a subquery. In Example 4, you could first use a subquery to find all part numbers in the ORDER_LINE table that appear in any row on which the order number is 21610. Then you could find the part description for any part whose part number is in this list. The query and its results appear in Figure 4.5.

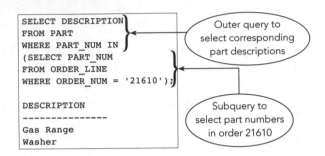

```
SELECT DESCRIPTION
FROM PART
WHERE PART_NUM IN
(SELECT PART_NUM
FROM ORDER_LINE
WHERE ORDER_NUM = '21610');

DESCRIPTION
----------------
Gas Range
Washer
```

Outer query to select corresponding part descriptions

Subquery to select part numbers in order 21610

In Figure 4.5, evaluating the subquery produces a temporary table consisting of those part numbers (DR93 and DW11) that are present in order number 21610. Executing the remaining portion of the query produces part descriptions for each part whose number is in this temporary table, in this case, Gas Range (DR93) and Washer (DW11).

M YSQL : The command shown in Figure 4.5 will not work in versions of MySQL prior
USER : to 4.1. In older versions, you would need to accomplish this task as a join.

Using EXISTS

You also can use the EXISTS operator to retrieve data from more than one table, as shown in Example 5.

E XAMPLE 5 : Find the order number and order date for each order that contains part number DR93.

This query is similar to the one in Example 4, but this time the query involves the ORDERS table and not the PART table. Here you can write the query in either of the ways previously demonstrated. For example, you could use the IN operator with a subquery, as shown in Figure 4.6. (Notice that the date is displayed as two digits, which is the default in some systems, even though you may have entered it into the table with four digits.)

FIGURE 4.6
....
Using the IN
operator to
select order
information

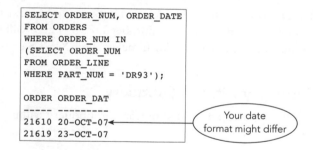

```
SELECT ORDER_NUM, ORDER_DATE
FROM ORDERS
WHERE ORDER_NUM IN
(SELECT ORDER_NUM
FROM ORDER_LINE
WHERE PART_NUM = 'DR93');


ORDER ORDER_DAT
----- ---------
21610 20-OCT-07
21619 23-OCT-07
```

Your date
format might differ

Using the EXISTS operator provides another approach, as shown in Figure 4.7.

FIGURE 4.7
....
Using the EXISTS
operator to
select order
information

```
SELECT ORDER_NUM, ORDER_DATE
FROM ORDERS
WHERE EXISTS
(SELECT *
FROM ORDER_LINE
WHERE ORDERS.ORDER_NUM = ORDER_LINE.ORDER_NUM
AND PART_NUM = 'DR93');

ORDER ORDER_DAT
----- ---------
21610 20-OCT-07
21619 23-OCT-07
```

M YSQL : The commands shown in Figures 4.6 and 4.7 will not work in versions of
  USER : MySQL prior to 4.1. In older versions you would need to accomplish these
       : tasks as a join.

The subquery in Figure 4.7 is the first one you have seen that involves a table listed in the outer query. This type of subquery is called a **correlated subquery**. In this case, the ORDERS table, which is listed in the FROM clause of the outer query, is used in the subquery. For this reason, you need to qualify the ORDER_NUM column in the subquery (ORDERS.ORDER_NUM). You did not need to qualify the columns in the previous queries involving the IN operator.

The query shown in Figure 4.7 works as follows. For each row in the ORDERS table, the subquery is executed using the value of ORDERS.ORDER_NUM that occurs in that row. The inner query produces a list of all rows in the ORDER_LINE table in which ORDER_LINE.ORDER_NUM matches this value and in which PART_NUM is equal to DR93. You can precede a subquery with the EXISTS operator to create a condition that is true if one or more rows are obtained when the subquery is executed; otherwise, the condition is false.

To illustrate the process, consider order numbers 21610 and 21613 in the ORDERS table. Order number 21610 is included because a row exists in the ORDER_LINE table with this order number and part number DR93. When the subquery is executed, there will be at least one row in the results, which in turn makes the EXISTS condition true. Order number 21613, however, will not be included because no row exists in the ORDER_LINE table with this order number and part number DR93. There will be no rows contained in the results of the subquery, which in turn makes the EXISTS condition false.

Using a Subquery within a Subquery

You can use SQL to create a **nested subquery** (a subquery within a subquery), as illustrated in Example 6.

EXAMPLE 6 : Find the order number and order date for each order that includes a part located in warehouse 3.

One way to approach this problem is first to determine the list of part numbers in the PART table for each part located in warehouse 3. Then you obtain a list of order numbers in the ORDER_LINE table with a corresponding part number in the part number list. Finally, you retrieve those order numbers and order dates in the ORDERS table for which the order number is in the list of order numbers obtained during the second step. The query and its results appear in Figure 4.8.

FIGURE 4.8
....
Nested
subqueries (a
subquery within
a subquery)

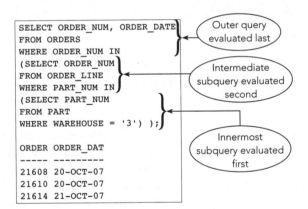

```
SELECT ORDER_NUM, ORDER_DATE
FROM ORDERS
WHERE ORDER_NUM IN
(SELECT ORDER_NUM
FROM ORDER_LINE
WHERE PART_NUM IN
(SELECT PART_NUM
FROM PART
WHERE WAREHOUSE = '3') );


ORDER ORDER_DAT
----- ---------
21608 20-OCT-07
21610 20-OCT-07
21614 21-OCT-07
```

Outer query evaluated last

Intermediate subquery evaluated second

Innermost subquery evaluated first

MYSQL : The command shown in Figure 4.8 will not work in versions of MySQL prior
USER : to 4.1. In older versions, you would need to accomplish this task as a join.

As you might expect, SQL evaluates the queries from the innermost query to the outermost query. The query in this example is evaluated in three steps:

1. The innermost subquery is evaluated first, producing a temporary table of part numbers for those parts located in warehouse 3.

2. The next (intermediate) subquery is evaluated, producing a second temporary table with a list of order numbers. Each order number in this collection has a row in the ORDER_LINE table for which the part number is in the temporary table produced in Step 1.

112

3. The outer query is evaluated last, producing the desired list of order numbers and order dates. Only those orders whose numbers are in the temporary table produced in Step 2 are included in the result.

Another approach involves joining the ORDERS, ORDER_LINE, and PART tables. The query and its results appear in Figure 4.9.

FIGURE 4.9

Joining three tables

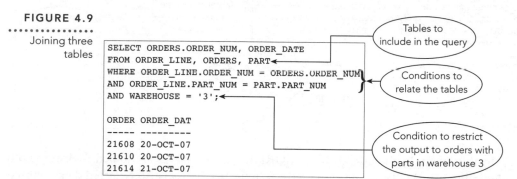

```
SELECT ORDERS.ORDER_NUM, ORDER_DATE
FROM ORDER_LINE, ORDERS, PART
WHERE ORDER_LINE.ORDER_NUM = ORDERS.ORDER_NUM
AND ORDER_LINE.PART_NUM = PART.PART_NUM
AND WAREHOUSE = '3';

ORDER ORDER_DAT
----- ---------
21608 20-OCT-07
21610 20-OCT-07
21614 21-OCT-07
```

Tables to include in the query

Conditions to relate the tables

Condition to restrict the output to orders with parts in warehouse 3

In this query, the conditions ORDER_LINE.ORDER_NUM = ORDERS.ORDER_NUM and ORDER_LINE.PART_NUM = PART.PART_NUM join the tables. The condition WAREHOUSE = '3' restricts the output to only those parts located in warehouse 3.

The query results are correct regardless of which formulation you use. You can use whichever approach you prefer.

You might wonder whether one approach is more efficient than the other. Many database management systems have built-in optimizers that analyze queries to determine the best way to satisfy them. Given a good optimizer, it should not make any difference how you formulate the query. If you are using a system without an optimizer, the formulation of a query *can* make a difference in the speed with which the query is executed. If you are working with a very large database and efficiency is a prime concern, consult your system's manual or try some timings yourself. Try running the same query both ways to see whether you notice a difference in the speed of execution. In small databases, there should not be a significant time difference between the two approaches.

A Comprehensive Example

The query used in Example 7 involves several of the features already discussed. It illustrates all the major clauses that you can use in the SELECT command. It also illustrates the order in which these clauses must appear.

EXAMPLE 7 : List the customer number, order number, order date, and order total for each order with a total that exceeds $1,000. Rename the order total as ORDER_TOTAL.

The query and its results appear in Figure 4.10.

FIGURE 4.10   Comprehensive example

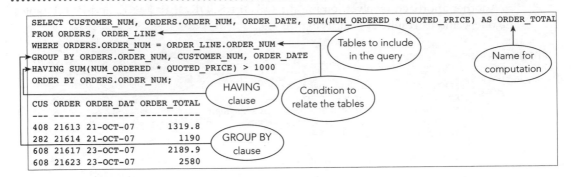

In this example, the ORDERS and ORDER_LINE tables are joined by listing both tables in the FROM clause and relating them in the WHERE clause. Selected data is sorted by order number using the ORDER BY clause. The GROUP BY clause indicates that the data is to be grouped by order number, customer number, and order date. For each group, the SELECT clause displays the customer number, order number, order date, and order total (SUM(NUM_ORDERED * QUOTED_PRICE)). In addition, the total was renamed ORDER_TOTAL. Not all groups will be displayed, however. The HAVING clause displays only those groups whose SUM(NUM_ORDERED * QUOTED_PRICE) is greater than $1,000.

The order number, customer number, and order date are unique for each order. Thus, it would seem that merely grouping by order number would be sufficient. Most SQL implementations require that both the customer number and the order date be listed in the GROUP BY clause. Recall that a SELECT clause can include statistics calculated for only the groups or columns whose values are identical for each row in a group. By stating that the data is to be grouped by order number, customer number, and order date, you tell the system that the values in these columns must be the same for each row in a group. A more sophisticated implementation would realize that, given the structure of this database, grouping by order number alone is sufficient to ensure the uniqueness of both customer number and order date.

Using an Alias

When tables are listed in the FROM clause, you can give each table an **alias**, or an alternate name, that you can use in the rest of the statement. You create an alias by typing the name of the table, pressing the Spacebar, and then typing the name of the alias. No commas or periods are necessary to separate the two names.

One reason for using an alias is simplicity. In Example 8, you will assign the REP table the alias R and the CUSTOMER table the alias C. By doing this, you can type R instead of REP and C instead of CUSTOMER in the remainder of the query. The query in this example is simple, so you might not see the full benefit of this feature. If the query is complex and requires you to qualify the names, using aliases can simplify the process.

EXAMPLE 8 : List the number, last name, and first name for each sales rep together with the number and name for each customer the sales rep represents.

The query and its results using aliases appear in Figure 4.11.

FIGURE 4.11 Using aliases

```
SELECT R.REP_NUM, LAST_NAME, FIRST_NAME, C.CUSTOMER_NUM, CUSTOMER_NAME
FROM REP R, CUSTOMER C
WHERE R.REP_NUM = C.REP_NUM;

RE LAST_NAME          FIRST_NAME          CUS CUSTOMER_NAME
-- ----------------   ----------------    --- ------------------------------
20 Kaiser             Valerie             148 Al's Appliance and Sport
35 Hull               Richard             282 Brookings Direct
65 Perez              Juan                356 Ferguson's
35 Hull               Richard             408 The Everything Shop
65 Perez              Juan                462 Bargains Galore
20 Kaiser             Valerie             524 Kline's
65 Perez              Juan                608 Johnson's Department Store
35 Hull               Richard             687 Lee's Sport and Appliance
35 Hull               Richard             725 Deerfield's Four Seasons
20 Kaiser             Valerie             842 All Season
```

Alias for the REP table

Alias for the CUSTOMER table

Note: Technically, it is unnecessary to qualify CUSTOMER_NUM because it is included only in the CUSTOMER table. It is qualified here for illustration purposes only.

Joining a Table to Itself

A second reason for using an alias is when you are joining a table to itself, called a **self-join,** as illustrated in Example 9.

EXAMPLE 9 : For each pair of customers located in the same city, display the customer number, customer name, and city.

If you had two separate tables for customers and the query requested customers in the first table having the same city as customers in the second table, you could use a normal join operation to find the answer. Here, however, there is only *one* table (CUSTOMER) that stores all the customer information. You can treat the CUSTOMER table as two tables in the query by creating an alias, as illustrated in Example 8. In this case, you would change the FROM clause to:

```
FROM CUSTOMER F, CUSTOMER S
```

SQL treats this clause as a query of two tables: one that has the alias F (first), and another that has the alias S (second). The fact that both tables are really the same CUSTOMER table is not a problem. The query and its results appear in Figure 4.12.

FIGURE 4.12   Using aliases for a self-join

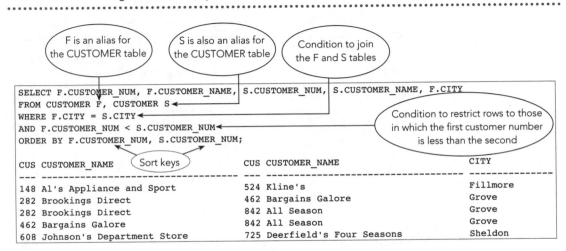

You are requesting a customer number and name from the F table, followed by a customer number and name from the S table, and then the city. (Because the city in the first table must match the city in the second, you can take the city from either table.) There are two conditions in the WHERE clause: The cities must match, and the customer number from the first table must be less than the customer number from the second table. In addition, the ORDER BY clause ensures that the data will be sorted by the first customer number. For those rows with the same first customer number, the data is further sorted by the second customer number.

Why is the condition F.CUSTOMER_NUM < S.CUSTOMER_NUM important in the query formulation?

ANSWER

If you did not include this condition, you'd get the query result shown in Figure 4.13. The first row is included because it is true that customer number 148 (Al's Appliance and Sport) in the F table has the same city as customer number 148 (Al's Appliance and Sport) in the S table. The second row indicates that customer number 148 (Al's Appliance and Sport) has the same city as customer number 524 (Kline's). The eleventh row, however, repeats the same information because customer number 524 (Kline's) has the same city as customer number 148 (Al's Appliance and Sport). Of these three rows, the only row that should be included in the query results is the second row. The second row is also the only one of the three rows in which the first customer number (148) is less than the second customer number (524). This is why the query requires the condition F.CUSTOMER_NUM < S.CUSTOMER_NUM.

FIGURE 4.13   Incorrect joining of a table to itself

```
SELECT F.CUSTOMER_NUM, F.CUSTOMER_NAME, S.CUSTOMER_NUM, S.CUSTOMER_NAME, F.CITY
FROM CUSTOMER F, CUSTOMER S
WHERE F.CITY = S.CITY
ORDER BY F.CUSTOMER_NUM, S.CUSTOMER_NUM;
```

First customer is the same as the second

Repeated information (both rows refer to the same pair of customers)

```
CUS CUSTOMER_NAME                    CUS CUSTOMER_NAME                         CITY
--- ------------------------------   --- ------------------------------       -----------
148 Al's Appliance and Sport         148 Al's Appliance and Sport             Fillmore
148 Al's Appliance and Sport         524 Kline's                              Fillmore
282 Brookings Direct                 282 Brookings Direct                     Grove
282 Brookings Direct                 462 Bargains Galore                      Grove
282 Brookings Direct                 842 All Season                           Grove
356 Ferguson's                       356 Ferguson's                           Northfield
408 The Everything Shop              408 The Everything Shop                  Crystal
462 Bargains Galore                  282 Brookings Direct                     Grove
462 Bargains Galore                  462 Bargains Galore                      Grove
462 Bargains Galore                  842 All Season                           Grove
524 Kline's                          148 Al's Appliance and Sport             Fillmore
524 Kline's                          524 Kline's                              Fillmore
608 Johnson's Department Store       608 Johnson's Department Store           Sheldon
608 Johnson's Department Store       725 Deerfield's Four Seasons             Sheldon
687 Lee's Sport and Appliance        687 Lee's Sport and Appliance            Altonville
725 Deerfield's Four Seasons         608 Johnson's Department Store           Sheldon
725 Deerfield's Four Seasons         725 Deerfield's Four Seasons             Sheldon
842 All Season                       282 Brookings Direct                     Grove
842 All Season                       462 Bargains Galore                      Grove
842 All Season                       842 All Season                           Grove
```

■ Using a Self-Join on a Primary Key

Occasionally a self-join might involve the primary key of a table. Figure 4.14, for example, shows some fields from an EMPLOYEE table whose primary key is EMPLOYEE_NUM. Another field in the table is MGR_EMPLOYEE_NUM, which represents the number of the employee's manager, who is also an employee. If you look at the row for employee 206 (Joan Dykstra), you will see she is managed by employee 198 (Mona Canzler). By looking at the row for employee 198 (Mona Canzler), you see that her manager is employee 108 (Martin Holden). In the row for employee 108 (Martin Holden), the manager number is null, indicating that he has no manager.

FIGURE 4.14
. . . .
Employee data

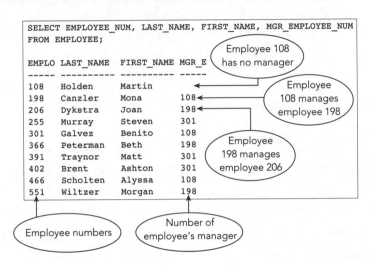

```
SELECT EMPLOYEE_NUM, LAST_NAME, FIRST_NAME, MGR_EMPLOYEE_NUM
FROM EMPLOYEE;


EMPLO LAST_NAME  FIRST_NAME MGR_E
----- ---------- ---------- -----
108   Holden     Martin
198   Canzler    Mona       108
206   Dykstra    Joan       198
255   Murray     Steven     301
301   Galvez     Benito     108
366   Peterman   Beth       198
391   Traynor    Matt       301
402   Brent      Ashton     301
466   Scholten   Alyssa     108
551   Wiltzer    Morgan     198
```

Employee 108 has no manager

Employee 108 manages employee 198

Employee 198 manages employee 206

Employee numbers

Number of employee's manager

Suppose you wanted to list the employee number, employee last name, and employee first name along with the number, last name, and first name of the employee's manager for each employee. Just as in the previous self-join, you would list the EMPLOYEE table twice in the FROM clause with aliases.

The command shown in Figure 4.15 uses the letter E as an alias for the employee and M as an alias for the manager. Thus E.EMPLOYEE_NUM would be the employee's number and M.EMPLOYEE_NUM would be the number for the employee's manager. In the SQL command, M.EMPLOYEE_NUM is renamed as MGR_NUM, M.LAST_NAME is renamed as MGR_LAST, and M.FIRST_NAME is renamed as MGR_FIRST. The condition in the WHERE clause ensures that E.MGR_EMPLOYEE_NUM (the number of the employee's manager) matches M.EMPLOYEE_NUM (the employee number on the manager's row in the table).

FIGURE 4.15 List of employees and their managers

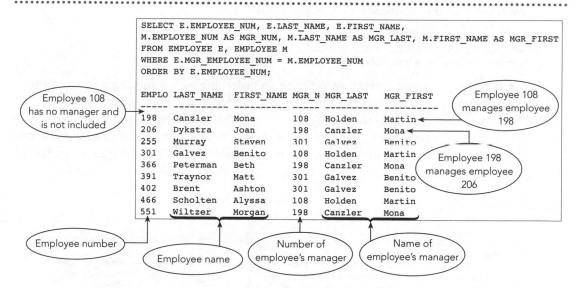

Joining Several Tables

It is possible to join several tables, as illustrated in Example 10. For each pair of tables you join, you must include a condition indicating how the columns are related.

EXAMPLE 10 : For each part on order, list the part number, number ordered, order number, order date, customer number, and customer name, along with the last name of the sales rep who represents each customer.

A part is on order if it occurs in any row in the ORDER_LINE table. The part number, number ordered, and order number appear in the ORDER_LINE table. If these requirements represent the entire query, you would write the query as follows:

```
SELECT PART_NUM, NUM_ORDERED, ORDER_NUM
FROM ORDER_LINE;
```

This formulation is not sufficient, however. You also need the order date and customer number, which are in the ORDERS table; the customer name, which is in the CUSTOMER table; and the rep last name, which is in the REP table. Thus, you need to join *four* tables: ORDER_LINE, ORDERS, CUSTOMER, and REP. The procedure for joining more than two tables is essentially the same as the one for joining two tables. The difference is that the condition in the WHERE clause will be a compound condition. In this case, you would write the WHERE clause as follows:

```
WHERE ORDERS.ORDER_NUM  = ORDER_LINE.ORDER_NUM
AND CUSTOMER.CUSTOMER_NUM = ORDERS.CUSTOMER_NUM
AND REP.REP_NUM = CUSTOMER.REP_NUM
```

The first condition relates an order to an order line with a matching order number. The second condition relates the customer to the order with a matching customer number. The final condition relates the rep to a customer with a matching sales rep number.

For the complete query, you list all the desired columns in the SELECT clause and qualify any columns that appear in more than one table. In the FROM clause, you list the tables that are involved in the query. The query and its results appear in Figure 4.16.

FIGURE 4.16 Joining four tables

```
SELECT PART_NUM, NUM_ORDERED, ORDER_LINE.ORDER_NUM, ORDER_DATE, CUSTOMER.CUSTOMER_NUM,
CUSTOMER_NAME, LAST_NAME
FROM ORDER_LINE, ORDERS, CUSTOMER, REP          Tables to include
WHERE ORDERS.ORDER_NUM = ORDER_LINE.ORDER_NUM   in the query
AND CUSTOMER.CUSTOMER_NUM = ORDERS.CUSTOMER_NUM   Conditions to
AND REP.REP_NUM = CUSTOMER.REP_NUM;               relate the tables

PART NUM_ORDERED ORDER ORDER_DAT CUS CUSTOMER_NAME                        LAST_NAME
---- ----------- ----- --------- --- ----------------------------------  ---------------
AT94          11 21608 20-OCT-07 148 Al's Appliance and Sport            Kaiser
DR93           1 21610 20-OCT-07 356 Ferguson's                         Perez
DW11           1 21610 20-OCT-07 356 Ferguson's                         Perez
KL62           4 21613 21-OCT-07 408 The Everything Shop                Hull
KT03           2 21614 21-OCT-07 282 Brookings Direct                   Hull
BV06           2 21617 23-OCT-07 608 Johnson's Department Store         Perez
CD52           4 21617 23-OCT-07 608 Johnson's Department Store         Perez
DR93           1 21619 23-OCT-07 148 Al's Appliance and Sport           Kaiser
KV29           2 21623 23-OCT-07 608 Johnson's Department Store         Perez
```

QUESTION   Why isn't the PART_NUM column, which appears in the PART and ORDER_LINE tables, qualified in the SELECT clause?

ANSWER   If the PART table appeared in the FROM clause, you would have to qualify PART_NUM; because this is not the case, the qualification is unnecessary. Among the tables listed in the query, only one table contains a column named PART_NUM.

The query shown in Figure 4.16 is more complex than many of the previous ones. You might think that SQL is not such an easy language to use after all. If you take it one step at a time, however, the query in Example 10 really isn't that difficult. To construct a detailed query in a step-by-step fashion, do the following:

1. List in the SELECT clause all the columns that you want to display. If the name of a column appears in more than one table, precede the column name with the table name (that is, qualify the column name).

2. List in the FROM clause all the tables involved in the query. Usually you include the tables that contain the columns listed in the SELECT clause. Occasionally, however, there might be a table that does not contain any columns used in the SELECT clause but that does contain columns used in the WHERE clause. In this case, you must also list the table in the FROM clause. For example, if you do not need to list a customer number or name, but you do need to list the rep name, you wouldn't include any columns from the CUSTOMER table in the SELECT clause. The CUSTOMER table is still required, however, because you must include columns from it in the WHERE clause.

3. Take one pair of related tables at a time and indicate in the WHERE clause the condition that relates the tables. Join these conditions with the AND operator. If there are any other conditions, include them in the WHERE clause and connect them to the other conditions with the AND operator. For example, if you wanted parts present on orders placed by only those customers with $10,000 credit limits, you would add one more condition to the WHERE clause, as shown in Figure 4.17.

FIGURE 4.17   Restricting the rows when joining four tables

```
SELECT PART_NUM, NUM_ORDERED, ORDER_LINE.ORDER_NUM, ORDER_DATE, CUSTOMER.CUSTOMER_NUM,
CUSTOMER_NAME, LAST_NAME
FROM ORDER_LINE, ORDERS, CUSTOMER, REP
WHERE ORDERS.ORDER_NUM = ORDER_LINE.ORDER_NUM
AND CUSTOMER.CUSTOMER_NUM = ORDERS.CUSTOMER_NUM
AND REP.REP_NUM = CUSTOMER.REP_NUM
AND CREDIT_LIMIT = 10000;


PART NUM_ORDERED ORDER ORDER_DAT CUS CUSTOMER_NAME                        LAST_NAME
---- ----------- ----- --------- --- ------------------------------------ ----------------
KT03           2 21614 21-OCT-07 282 Brookings Direct                     Hull
BV06           2 21617 23-OCT-07 608 Johnson's Department Store           Perez
CD52           4 21617 23-OCT-07 608 Johnson's Department Store           Perez
KV29           2 21623 23-OCT-07 608 Johnson's Department Store           Perez
```

■ Set Operations

In SQL, you can use the set operations for taking the union, intersection, and difference of two tables. The **union** of two tables is a table containing every row that is in either the first table, the second table, or both tables. The **intersection** (**intersect**) of two tables is a table containing all rows that are in both tables. The **difference (minus)** of two tables is the set of all rows that are in the first table but that are not in the second table.

For example, suppose that TEMP1 is a table containing the number and name of each customer represented by sales rep 65. Further suppose that TEMP2 is a table containing the number and name of those customers that currently have orders on file, as shown in Figure 4.18.

FIGURE 4.18   Customers of rep 65 and customers with open orders

TEMP1

CUSTOMER_NUM	CUSTOMER_NAME
356	Ferguson's
462	Bargains Galore
608	Johnson's Department Store

TEMP2

CUSTOMER_NUM	CUSTOMER_NAME
148	Al's Appliance and Sport
282	Brookings Direct
356	Ferguson's
408	The Everything Shop
608	Johnson's Department Store

The union of TEMP1 and TEMP2 (TEMP1 UNION TEMP2) consists of the number and name of those customers that are represented by sales rep 65 *or* that currently have orders on file, *or* both. The intersection of these two tables (TEMP1 INTERSECT TEMP2) contains those customers that are represented by sales rep 65 *and* that have orders on file. The difference of these two tables (TEMP1 MINUS TEMP2) contains those customers that are represented by sales rep 65 but that *do not* have orders on file. The results of these set operations are shown in Figure 4.19.

FIGURE 4.19   Union, intersect, and minus of the TEMP1 and TEMP2 tables

TEMP1 UNION TEMP2

CUSTOMER_NUM	CUSTOMER_NAME
148	Al's Appliance and Sport
282	Brookings Direct
356	Ferguson's
408	The Everything Shop
462	Bargains Galore
608	Johnson's Department Store

TEMP1 INTERSECT TEMP2

CUSTOMER_NUM	CUSTOMER_NAME
356	Ferguson's
608	Johnson's Department Store

TEMP1 MINUS TEMP2

CUSTOMER_NUM	CUSTOMER_NAME
462	Bargains Galore

There is a restriction on set operations. It does not make sense, for example, to talk about the union of the CUSTOMER table and the ORDERS table because these tables don't contain the same columns. What might rows in this union look like? The two tables in the union *must* have the same structure for a union to be appropriate; the formal term is "union compatible." Two tables are **union compatible** if they have the same number of columns and if their corresponding columns have identical data types and lengths.

Note that the definition of union compatible does not state that the columns of the two tables must be identical but rather that the columns must be of the same type. Thus, if one column is CHAR(20), the matching column also must be CHAR(20).

EXAMPLE 11 : List the number and name of each customer that either is represented by sales rep 65 or that currently has orders on file, or both.

You can create a table containing the number and name of each customer that is represented by sales rep 65 by selecting the customer numbers and names from the CUSTOMER table for which the sales rep number is 65. Then you can create another table containing the number and name of each customer that currently has orders on file by creating a join of the CUSTOMER and ORDERS tables. The two tables created by this process have the same structure; that is, they both contain the CUSTOMER_NUM and CUSTOMER_NAME columns. Because the tables are union compatible, it is possible to take the union of these two tables. The query and its results appear in Figure 4.20.

FIGURE 4.20
....
Using the UNION
operator

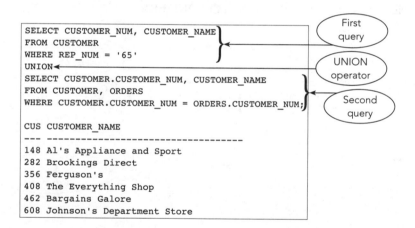

```
SELECT CUSTOMER_NUM, CUSTOMER_NAME
FROM CUSTOMER
WHERE REP_NUM = '65'
UNION
SELECT CUSTOMER.CUSTOMER_NUM, CUSTOMER_NAME
FROM CUSTOMER, ORDERS
WHERE CUSTOMER.CUSTOMER_NUM = ORDERS.CUSTOMER_NUM;

CUS CUSTOMER_NAME
--- -----------------------------------
148 Al's Appliance and Sport
282 Brookings Direct
356 Ferguson's
408 The Everything Shop
462 Bargains Galore
608 Johnson's Department Store
```

If the SQL implementation truly supports the union operation, it will remove any duplicate rows automatically. For example, any customer that is represented by sales rep 65 *and* that currently has orders on file will appear only once in the results. Oracle, Access, and MySQL correctly remove duplicates. Other SQL implementations, however, support a union operation but do not remove such duplicates.

EXAMPLE 12 : List the number and name of each customer that is represented by sales rep 65 and that currently has orders on file.

The only difference between this query and the one in Example 11 is that here, the appropriate operator is INTERSECT. The query and its results appear in Figure 4.21.

FIGURE 4.21
....
Using the
INTERSECT
operator

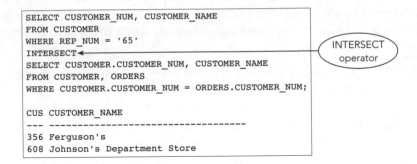

```
SELECT CUSTOMER_NUM, CUSTOMER_NAME
FROM CUSTOMER
WHERE REP_NUM = '65'
INTERSECT
SELECT CUSTOMER.CUSTOMER_NUM, CUSTOMER_NAME
FROM CUSTOMER, ORDERS
WHERE CUSTOMER.CUSTOMER_NUM = ORDERS.CUSTOMER_NUM;


CUS CUSTOMER_NAME
--- --------------------------------
356 Ferguson's
608 Johnson's Department Store
```

INTERSECT
operator

Many SQL implementations, including Access and MySQL, do not support the INTERSECT operator. In such a case, you need to take a different approach, such as the one shown in Figure 4.22. This command produces the same results as using the INTERSECT operator. The command selects the number and name of each customer that is represented by sales rep 65 and whose customer number also appears in the collection of customer numbers in the ORDERS table.

FIGURE 4.22
....
Performing
an intersection
without using the
INTERSECT
operator

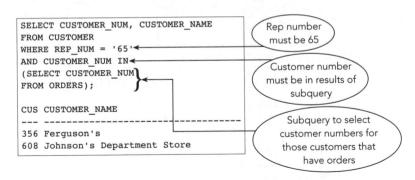

```
SELECT CUSTOMER_NUM, CUSTOMER_NAME
FROM CUSTOMER
WHERE REP_NUM = '65'
AND CUSTOMER_NUM IN
(SELECT CUSTOMER_NUM
FROM ORDERS);

CUS CUSTOMER_NAME
--- --------------------------------
356 Ferguson's
608 Johnson's Department Store
```

Rep number
must be 65

Customer number
must be in results of
subquery

Subquery to select
customer numbers for
those customers that
have orders

MYSQL : The command shown in Figure 4.22 will not work in versions of MySQL
USER : prior to 4.1. In older versions, you would need to accomplish this task as
: a join.

EXAMPLE 13 : List the number and name of each customer that is represented by sales
: rep 65 but that does not have orders currently on file.

The only difference between this query and the ones in Examples 11 and 12 is that here, the appropriate operator is MINUS. The query and its results appear in Figure 4.23.

FIGURE 4.23
. . . .
Using the MINUS
operator

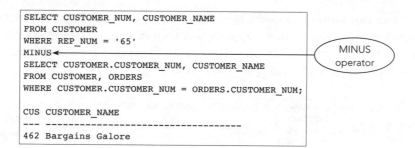

```
SELECT CUSTOMER_NUM, CUSTOMER_NAME
FROM CUSTOMER
WHERE REP_NUM = '65'
MINUS◄
SELECT CUSTOMER.CUSTOMER_NUM, CUSTOMER_NAME
FROM CUSTOMER, ORDERS
WHERE CUSTOMER.CUSTOMER_NUM = ORDERS.CUSTOMER_NUM;

CUS CUSTOMER_NAME
--- --------------------------------
462 Bargains Galore
```

MINUS
operator

SQL implementations (including Access and MySQL) that do not support the INTERSECT operator usually do not support the MINUS operator either. In such cases, you need to take a different approach, such as the one shown in Figure 4.24. This command produces the same results by selecting the number and name of each customer that is represented by sales rep 65 and whose customer number does *not* appear in the collection of customer numbers in the ORDERS table.

FIGURE 4.24
. . . .
Performing a
difference
without using the
MINUS operator

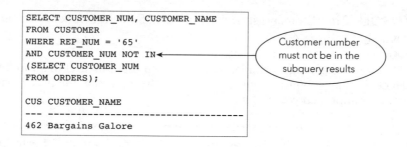

```
SELECT CUSTOMER_NUM, CUSTOMER_NAME
FROM CUSTOMER
WHERE REP_NUM = '65'
AND CUSTOMER_NUM NOT IN◄
(SELECT CUSTOMER_NUM
FROM ORDERS);

CUS CUSTOMER_NAME
--- --------------------------------
462 Bargains Galore
```

Customer number
must not be in the
subquery results

M**YSQL** : The command shown in Figure 4.24 will not work in versions of MySQL
USER : prior to 4.1.

■ ALL and ANY

You can use the ALL and ANY operators with subqueries to produce a single column of numbers. If you precede the subquery by the ALL operator, the condition is true only if it satisfies *all* values produced by the subquery. If you precede the subquery by the ANY operator, the condition is true only if it satisfies *any* value (one or more) produced by the subquery. The next examples illustrate the use of these operators.

E**XAMPLE 14** : Find the customer number, name, current balance, and rep number of
each customer whose balance is greater than the individual balances of
each customer of sales rep 65.

You can satisfy this query by finding the maximum balance of the customers represented by sales rep 65 in a subquery and then finding all customers whose balances are greater than this number; however, there is an alternative. You can use the ALL operator, as shown in Figure 4.25, to simplify the process.

FIGURE 4.25
· · · · · · · · · · · · · · ·
SELECT
command with
an ALL condition

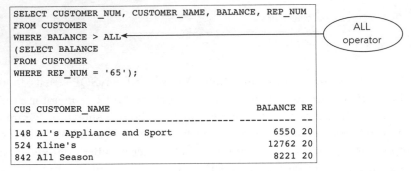

```
SELECT CUSTOMER_NUM, CUSTOMER_NAME, BALANCE, REP_NUM
FROM CUSTOMER
WHERE BALANCE > ALL
(SELECT BALANCE
FROM CUSTOMER
WHERE REP_NUM = '65');


CUS CUSTOMER_NAME                          BALANCE RE
--- ------------------------------------- ---------- --
148 Al's Appliance and Sport                 6550 20
524 Kline's                                  12762 20
842 All Season                                8221 20
```

ALL
operator

MYSQL : The command shown in Figure 4.25 will not work in versions of MySQL
USER : prior to 4.1.

To some users, this formulation might seem more natural than finding the maximum balance in the subquery. For other users, the opposite might be true. You can use whichever approach you prefer.

QUESTION   How would you get the same result for Example 14 without using the ALL operator?

ANSWER   You could select each customer whose balance is greater than the maximum balance of any customer of sales rep 65, as shown in Figure 4.26.

FIGURE 4.26
· · · · · · · · · · · · · · ·
Alternative to
using an ALL
condition

```
SELECT CUSTOMER_NUM, CUSTOMER_NAME, BALANCE, REP_NUM
FROM CUSTOMER
WHERE BALANCE >
(SELECT MAX(BALANCE)
FROM CUSTOMER
WHERE REP_NUM = '65');


CUS CUSTOMER_NAME                          BALANCE RE
--- ------------------------------------- ---------- --
148 Al's Appliance and Sport                 6550 20
524 Kline's                                  12762 20
842 All Season                                8221 20
```

The command shown in Figure 4.26 will not work in versions of MySQL prior to 4.1. In older versions, you would need to accomplish this task in two steps. First, you would use a query to find the largest balance for a customer of rep 65. Then you would use this result in a second query to find all customers with balances that are more than the maximum balance you determined in the first query.

EXAMPLE 15

Find the customer number, name, current balance, and rep number of each customer whose balance is greater than the balance of at least one customer of sales rep 65.

You can satisfy this query by finding the minimum balance of the customers represented by sales rep 65 in a subquery and then finding all customers whose balance is greater than this number. To simplify the process, you can use the ANY operator, as shown in Figure 4.27.

FIGURE 4.27

SELECT command with an ANY condition

```
SELECT CUSTOMER_NUM, CUSTOMER_NAME, BALANCE, REP_NUM
FROM CUSTOMER
WHERE BALANCE > ANY←                                    ANY
(SELECT BALANCE                                         operator
FROM CUSTOMER
WHERE REP_NUM = '65');


CUS CUSTOMER_NAME                           BALANCE RE
--- ------------------------------------ ---------- --
148 Al's Appliance and Sport                  6550 20
356 Ferguson's                                5785 65
408 The Everything Shop                    5285.25 35
462 Bargains Galore                           3412 65
524 Kline's                                  12762 20
687 Lee's Sport and Appliance                 2851 35
842 All Season                                8221 20
```

The command shown in Figure 4.27 will not work in versions of MySQL prior to 4.1.

QUESTION

ANSWER

How would you get the same results without using the ANY operator?

You could select each customer whose balance is greater than the minimum balance of any customer of sales rep 65, as shown in Figure 4.28.

```
SELECT CUSTOMER_NUM, CUSTOMER_NAME, BALANCE, REP_NUM
FROM CUSTOMER
WHERE BALANCE >
(SELECT MIN(BALANCE)
FROM CUSTOMER
WHERE REP_NUM = '65');


CUS CUSTOMER_NAME                           BALANCE RE
--- ------------------------------------ ---------- --
148 Al's Appliance and Sport                  6550 20
356 Ferguson's                                5785 65
408 The Everything Shop                     5285.25 35
462 Bargains Galore                            3412 65
524 Kline's                                   12762 20
687 Lee's Sport and Appliance                 2851 35
842 All Season                                 8221 20
```

MYSQL
USER

The command shown in Figure 4.28 will not work in versions of MySQL prior to 4.1. In older versions, you would need to accomplish this task in two steps. First, you would use a query to find the smallest balance for a customer of rep 65. Then you would use this result in a second query to find all customers with balances that are more than the minimum balance you determined in the first query.

■ Special Operations

You can perform special operations within SQL, such as the self-join that you already used. Three other special operations are the inner join, the outer join, and the product.

Inner Join

A join that compares the tables in the FROM clause and lists only those rows that satisfy the condition in the WHERE clause is called an **inner join**. The joins that you have performed so far in this text have been inner joins. Example 16 explains the inner join.

EXAMPLE 16 : Display the customer number, customer name, order number, and order date for each order. Sort the results by customer number.

This example requires the same type of join that you have been using. The command is:

```
SELECT CUSTOMER.CUSTOMER_NUM, CUSTOMER_NAME, ORDER_NUM, ORDER_DATE
FROM CUSTOMER, ORDERS
WHERE CUSTOMER.CUSTOMER_NUM = ORDERS.CUSTOMER_NUM
ORDER BY CUSTOMER.CUSTOMER_NUM;
```

The previous approach should work in any SQL implementation. An update to the SQL standard approved in 1992, called SQL-92, provides an alternative way of performing an inner join, as demonstrated in Figure 4.29. Many SQL implementations support at least a portion of the SQL-92 standard.

FIGURE 4.29 JOIN example

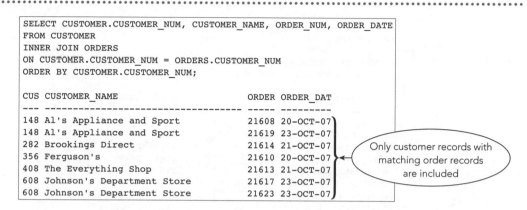

```
SELECT CUSTOMER.CUSTOMER_NUM, CUSTOMER_NAME, ORDER_NUM, ORDER_DATE
FROM CUSTOMER
INNER JOIN ORDERS
ON CUSTOMER.CUSTOMER_NUM = ORDERS.CUSTOMER_NUM
ORDER BY CUSTOMER.CUSTOMER_NUM;

CUS CUSTOMER_NAME                          ORDER ORDER_DAT
--- ----------------------------------    ----- ---------
148 Al's Appliance and Sport              21608 20-OCT-07
148 Al's Appliance and Sport              21619 23-OCT-07
282 Brookings Direct                      21614 21-OCT-07
356 Ferguson's                            21610 20-OCT-07
408 The Everything Shop                   21613 21-OCT-07
608 Johnson's Department Store            21617 23-OCT-07
608 Johnson's Department Store            21623 23-OCT-07
```

Only customer records with matching order records are included

In the FROM clause, list the first table, then include the words INNER JOIN followed by the name of the other table. Instead of a WHERE clause, use an ON clause containing the same condition that you would have included in the WHERE clause.

Outer Join

Sometimes you need to list all the rows from one of the tables in a join, regardless of whether they match any rows in the other table. For example, you can perform the join of the CUSTOMER and ORDERS tables in the query for Example 16, but display all customers—even the ones without orders. This type of join is called an **outer join**.

There are actually three types of outer joins. In a **left outer join**, all rows from the table on the left (the table listed first in the query) will be included regardless of whether they match rows from the table on the right (the table listed second in the query). Rows from the table on the right will be included only if they match. In a **right outer join**, all rows from the table on the right will be included regardless of whether they match rows from the table on the left. Rows from the table on the left will be included only if they match. In a **full outer join**, all rows from both tables will be included regardless of whether they match rows from the other table. (The full outer join is rarely used.)

EXAMPLE 17 : Display the customer number, customer name, order number, and order date for all orders. Include all customers in the results. For customers that do not have orders, omit the order number and order date.

To include all customers, you must perform an outer join. Assuming the CUSTOMER table is listed first, the join should be a left outer join. In SQL, you use the LEFT JOIN clause to perform a left outer join as shown in Figure 4.30. (You would use a RIGHT JOIN clause to perform a right outer join.)

FIGURE 4.30 Outer join example

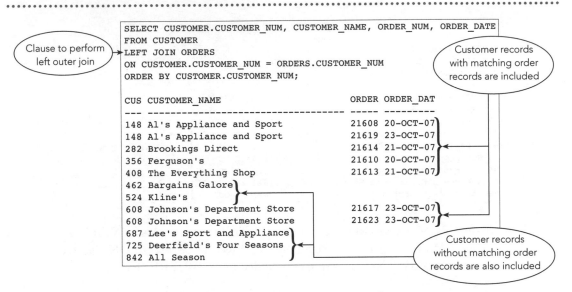

All customers are included in the results. For customers without orders, the order number and date are blank. Technically, these blank values are null.

ORACLE USER   In Oracle, there is another way to perform left and right outer joins. You write the join as you have been doing, with one exception. You include parentheses and a plus sign in the WHERE clause after the column in the table for which only matching rows are to be included. In this example, the plus sign would follow the CUSTOMER_NUM column in the ORDERS table because only orders that match customers are to be included. Because customers that do not have orders are to be included in the results, there would be no plus sign after the CUSTOMER_NUM column in the CUSTOMER table. The correct query is as follows:

```
SELECT CUSTOMER.CUSTOMER_NUM, CUSTOMER_NAME, ORDER_NUM, ORDER_DATE
FROM CUSTOMER, ORDERS
WHERE CUSTOMER.CUSTOMER_NUM = ORDERS.CUSTOMER_NUM(+)
ORDER BY CUSTOMER.CUSTOMER_NUM;
```

Running this query produces the same results shown in Figure 4.30.

Product

The **product** (formally called the **Cartesian Product**) of two tables is the combination of all rows in the first table and all rows in the second table.

Note: This operation is not common. You need to be aware of it, however, because it is easy to create a product inadvertently by omitting the WHERE clause when you are attempting to join tables.

EXAMPLE 18 : Form the product of the CUSTOMER and ORDERS tables. Display the customer number and name from the CUSTOMER table, along with the order number and order date from the ORDERS table.

Forming a product is actually very easy. You simply omit the WHERE clause, as shown in Figure 4.31.

FIGURE 4.31 Product example

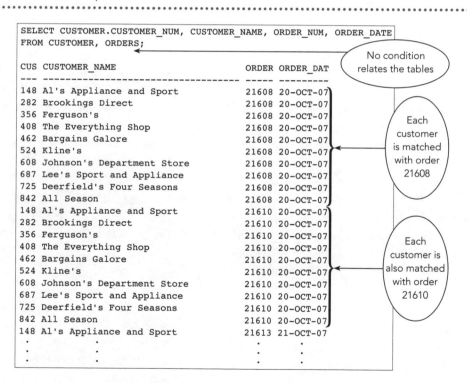

In this chapter, you learned how to use SQL commands to join tables. You qualified column names when necessary to obtain the desired results. You used subqueries to query multiple tables and used the IN and EXISTS operators. You used an alias and examined situations in which using aliases is beneficial. You used the SQL set operators UNION, INTERSECT, and MINUS to find rows in one, two, or both tables. You used the ALL and ANY operators to simplify certain SQL queries. You learned how to perform inner joins, outer joins, and products. In the next chapter, you will use SQL commands to update the data in your tables.

■ SUMMARY

- To join tables, indicate in the SELECT clause all columns to display, list in the FROM clause all tables to join, and then include in the WHERE clause any conditions requiring values in matching columns to be equal.

- When referring to matching columns in different tables, you must qualify the column names to avoid confusion. You qualify column names using the following format: table name.column name.

- Use the IN or EXISTS operators with an appropriate subquery as an alternate way of performing a join.

- A subquery can contain another subquery. The innermost subquery is executed first.

- The name of a table in a FROM clause can be followed by an alias, which is an alternate name for a table. The alias can be used in place of the table name throughout the SQL command.

- By using two different aliases for the same table in a single SQL command, you can join a table to itself.

- The UNION command creates a union of two tables (the collection of rows that are in either or both tables). The INTERSECT command creates the intersection of two tables (the collection of rows that are in both tables). The MINUS command creates the difference of two tables (the collection of rows that are in the first table but not in the second table). To perform any of these operations, the tables must be union compatible.

- Two tables are union compatible if they have the same number of columns and their corresponding columns have identical data types and lengths.

- If the ALL operator precedes a subquery, the condition is true only if it is satisfied by *all* values produced by the subquery.

- If the ANY operator precedes a subquery, the condition is true only if it is satisfied by *any* value (one or more) produced by the subquery.

- In an inner join, only matching rows from both tables are included. You can use the INNER JOIN clause to perform an inner join.

- In a left outer join, all rows from the table on the left (the table listed first in the query) will be included regardless of whether they match rows from the table on the right (the table listed second in the query). Rows from the table on the right will be included only if they match. You can use the LEFT JOIN clause to perform a left outer join. In a right outer join, all rows from the table on the right will be included regardless of whether they match rows from the table on the left. Rows from the table on the left will be included only if they match. You can use the RIGHT JOIN clause to perform a right outer join.

- The product (Cartesian Product) of two tables is the combination of all rows in the first table and all rows in the second table. To form a product of two tables, include both tables in the FROM clause and omit the WHERE clause.

■ KEY TERMS

alias	left outer join
Cartesian Product	minus
correlated subquery	nested subquery
difference	outer join
full outer join	product
inner join	right outer join
intersect	self-join
intersection	union
join	union compatible

■ REVIEW QUESTIONS

1. How do you join tables in SQL?

2. When must you qualify names in SQL commands? How do you do so?

3. List two operators that you can use with subqueries as an alternate way of performing joins.

4. What is a nested subquery? In which order does SQL evaluate nested subqueries?

5. What is an alias? How do you specify one in SQL? How do you use an alias?

6. How do you join a table to itself in SQL?

7. How do you take the union of two tables in SQL? How do you take the intersection of two tables? How do you take the difference of two tables? Are there any restrictions on the tables when performing any of these operations?

8. What does it mean for two tables to be union compatible?

9. How do you use the ALL operator with a subquery?

10. How do you use the ANY operator with a subquery?

11. Which rows are included in an inner join? What clause can you use to perform an inner join in SQL?

12. Which rows are included in a left outer join? What clause can you use to perform a left outer join in SQL?

13. Which rows are included in a right outer join? What clause can you use to perform a right outer join in SQL?

14. What is the formal name for the product of two tables? How do you form a product in SQL?

■ EXERCISES (Premiere Products)

Use SQL and the Premiere Products database (see Figure 1.2 in Chapter 1) to complete the following exercises. Use the notes at the end of Chapter 2 to print your output if directed to do so by your instructor.

1. For each order, list the order number and order date along with the number and name of the customer that placed the order.

2. For each order placed on October 21, 2007, list the order number along with the number and name of the customer that placed the order.

3. For each order, list the order number, order date, part number, number of units ordered, and quoted price for each order line that makes up the order.

4. If your SQL implementation supports the IN operator, use the IN operator to find the number and name of each customer that placed an order on October 21, 2007.

5. If your SQL implementation supports the EXISTS operator, repeat Exercise 4, but this time use the EXISTS operator in your answer.

6. Find the number and name of each customer that did not place an order on October 21, 2007.

7. For each order, list the order number, order date, part number, part description, and item class for each part that makes up the order.

8. Repeat Exercise 7, but this time order the rows by item class and then by order number.

9. Use a subquery to find the rep number, last name, and first name of each sales rep who represents at least one customer with a credit limit of $5,000. List each sales rep only once in the results.

10. Repeat Exercise 9, but this time do not use a subquery.

11. Find the number and name of each customer that currently has an order on file for a Gas Range.

12. List the part number, part description, and item class for each pair of parts that are in the same item class. (For example, one such pair would be part AT94 and part FD21, because the item class for both parts is HW.)

13. List the order number and order date for each order placed by the customer named Johnson's Department Store. (*Hint:* To enter an apostrophe (single quotation mark) within a string of characters, type two single quotation marks.)

14. List the order number and order date for each order that contains an order line for an Iron.

15. List the order number and order date for each order that was either placed by Johnson's Department Store or that contains an order line for a Gas Range.

16. List the order number and order date for each order that was placed by Johnson's Department Store and that contains an order line for a Gas Range.

17. List the order number and order date for each order that was placed by Johnson's Department Store but that does not contain an order line for a Gas Range.

18. List the part number, part description, unit price, and item class for each part that has a unit price greater than the unit price of every part in item class AP. Use either the ALL or ANY operator in your query. (*Hint:* Make sure you select the correct operator.)

19. If you used ALL in Exercise 18, repeat the exercise using ANY. If you used ANY, repeat the exercise using ALL, then run the new command. What question does this command answer?

20. For each part, list the part number, description, units on hand, order number, and number of units ordered. All parts should be included in the results. For those parts that are currently not on order, the order number and number of units ordered should be left blank. Order the results by part number.

■ EXERCISES (Henry Books)

Use SQL and the Henry Books database (Figures 1.4 through 1.7 in Chapter 1) to complete the following exercises. Use the Notes at the end of Chapter 2 to print your output if directed to do so by your instructor.

1. For each book, list the book code, book title, publisher code, and publisher name. Order the results by publisher name.

2. For each book published by Plume, list the book code, book title, and price.

3. List the book title, book code, and price of each book published by Plume that has a book price of at least $14.

4. List the book code, book title, and units on hand for each book in branch number 4.

5. List the book title for each book that has the type PSY and that is published by Jove Publications.

6. Find the book title for each book written by author number 18. If your SQL implementation supports the IN operator, use the IN operator in your formulation.

7. If your SQL implementation supports the EXISTS operator, repeat Exercise 6, but this time use the EXISTS operator in your formulation.

8. Find the book code and book title for each book located in branch number 2 and written by author 20.

9. List the book codes for each pair of books that have the same price. (For example, one such pair would be book 0200 and book 7559, because the price of both books is $8.00.) The first book code listed should

be the major sort key and the second book code should be the minor sort key.

10. Find the book title, author last name, and units on hand for each book in branch number 4.

11. Repeat Exercise 10, but this time list only paperback books.

12. Find the book code and book title for each book whose price is more than $10 or that was published in Boston.

13. Find the book code and book title for each book whose price is more than $10 and that was published in Boston.

14. Find the book code and book title for each book whose price is more than $10 but that was not published in Boston.

15. Find the book code and book title for each book whose price is greater than the book price of every book that has the type HOR.

16. Find the book code and book title for each book whose price is greater than the price of at least one book that has the type HOR.

17. List the book code, book title, and units on hand for each book in branch number 2. Be sure each book is included, regardless of whether there are any copies of the book currently on hand in branch 2. Order the output by book code.

▪EXERCISES (Alexamara Marina Group)

Use SQL and the Alexamara Marina Group database (Figures 1.8 through 1.12 in Chapter 1) to complete the following exercises. Use the Notes at the end of Chapter 2 to print your output if directed to do so by your instructor.

1. For every boat, list the marina number, slip number, boat name, owner number, owner's first name, and owner's last name.

2. For every completed or open service request for routine engine maintenance, list the slip ID, description, and status.

3. For every service request for routine engine maintenance, list the slip ID, marina number, slip number, estimated hours, spent hours, owner number, and owner's last name.

4. List the first and last names of all owners who have a boat in a 40-foot slip. If your SQL implementation supports the IN operator, use the IN operator in your formulation.

5. If your SQL implementation supports the EXISTS operator, repeat Exercise 4, but this time use the EXISTS operator in your formulation.

6. List the names of any pair of boats that have the same type. For example, one pair would be *Anderson II* and *Escape*, because the boat type for both boats is Sprite 4000. The first name listed should be the major sort key and the second name should be the minor sort key.

7. List the boat name, owner number, owner last name, and owner first name for each boat in marina 1.

8. Repeat Exercise 7, but this time only list boats in 30-foot slips.

9. List the marina number, slip number, and boat name for boats whose owner lives in Glander Bay or whose type is Sprite 4000.

10. List the marina number, slip number, and boat name for boats whose owner lives in Glander Bay and whose type is Sprite 4000.

11. List the marina number, slip number, and boat name for boats whose owner lives in Glander Bay but whose type is not Sprite 4000.

12. Find the service ID and slip ID for each service request whose estimated hours is greater than the number of estimated hours of at least one service request on which the category number is 3.

13. Find the service ID and slip ID for each service request whose estimated hours is greater than the number of estimated hours on every service request on which the category number is 3.

14. List the slip ID, boat name, owner number, service ID, number of estimated hours, and number of spent hours for each service request on which the category number is 2.

15. Repeat Exercise 14, but this time be sure each slip is included regardless of whether the boat in the slip currently has any service requests for category 2.

CHAPTER 5

Updating Data

OBJECTIVES

- Create a new table from an existing table

- Change data using the UPDATE command

- Add new data using the INSERT command

- Use the COMMIT and ROLLBACK commands to make permanent data updates or to reverse updates

- Understand transactions and the role of COMMIT and ROLLBACK in supporting transactions

- Delete data using the DELETE command

- Use nulls in UPDATE commands

- Change the structure of an existing table

- Drop a table

Introduction

In this chapter, you will learn how to create a new table from an existing table and make changes to the data in a table. You will use the UPDATE command to change data in one or more rows in a table, and use the INSERT command to add new rows. You will learn how to use the COMMIT and ROLLBACK commands to make changes permanent and return changes to their original state. You will use the DELETE command to delete rows. You will also use nulls in update operations. Finally, you will learn how to change the structure of a table in a variety of ways and drop existing tables.

■ Creating a New Table from an Existing Table

It is possible to create a new table from data in an existing table, as illustrated in the following examples.

EXAMPLE 1   Create a new table named LEVEL1_CUSTOMER containing the following columns from the CUSTOMER table: CUSTOMER_NUM, CUSTOMER_NAME, BALANCE, CREDIT_LIMIT, and REP_NUM. The columns in the new LEVEL1_CUSTOMER table should have the same characteristics as the corresponding columns in the CUSTOMER table.

You describe the new table named LEVEL1_CUSTOMER by using the CREATE TABLE command shown in Figure 5.1.

FIGURE 5.1   Creating the LEVEL1_CUSTOMER table

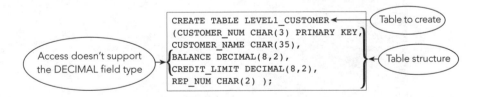

```
CREATE TABLE LEVEL1_CUSTOMER          ◄──── Table to create
(CUSTOMER_NUM CHAR(3) PRIMARY KEY,
CUSTOMER_NAME CHAR(35),
BALANCE DECIMAL(8,2),
CREDIT_LIMIT DECIMAL(8,2),
REP_NUM CHAR(2) );
```

Access doesn't support the DECIMAL field type

Table structure

ACCESS USER   In Access, use either the CURRENCY (for dollar amounts) or NUMBER (for other numeric fields) field types in place of the DECIMAL field type. Unlike the DECIMAL field type, you do not enter the field size and number of decimal places. Rather than DECIMAL (8,2) for example, you would use the CURRENCY field type for the BALANCE field, which contains a dollar amount.

EXAMPLE 2 : Insert into the LEVEL1_CUSTOMER table the customer number, customer name, balance, credit limit, and rep number for customers with credit limits of $7,500.

You can create a SELECT command to select the desired data from the CUSTOMER table, just as you did in Chapter 3. By placing this SELECT command in an INSERT command, you can add the query results to a table. The INSERT command appears in Figure 5.2; this command inserts four rows into the LEVEL1_CUSTOMER table.

FIGURE 5.2 Inserting data into the LEVEL1_CUSTOMER table

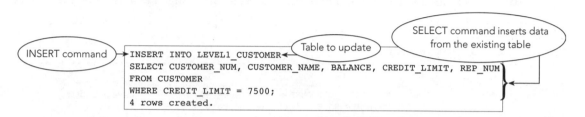

The SELECT command shown in Figure 5.3 displays the data in the LEVEL1_CUSTOMER table. Notice that the data comes from the new table you just created (LEVEL1_CUSTOMER), not the CUSTOMER table.

FIGURE 5.3 LEVEL1_CUSTOMER data

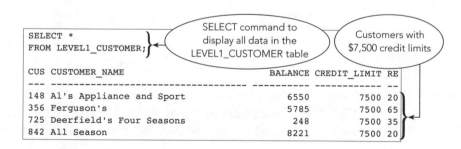

■ Changing Existing Data in a Table

The data stored in your tables is subject to constant change; prices, addresses, commission amounts, and other data in a database change on a regular basis. To keep data current, you must be able to make these changes to the data in your tables. You can use the **UPDATE** command to change rows on which a specific condition is true.

: Change the name of customer 842 in the LEVEL1_CUSTOMER table to All
Season Sport.

The format for the UPDATE command is the word UPDATE, followed by the name
of the table to be updated. The next portion of the command consists of the word **SET**, followed by the name of the column to be updated, an equals sign, and the new value. When
necessary, include a WHERE clause to indicate the row(s) on which the change is to take
place. The command shown in Figure 5.4 changes the name of customer 842 to All Season
Sport. The SELECT command shown in Figure 5.4 shows the data in the table after the
change has been made. (This SELECT command is not part of the update.) It is a good
idea to use a SELECT command to display the data you changed to verify that the correct
change was made.

FIGURE 5.4   Updating a row in a table

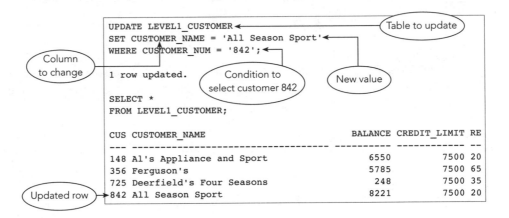

: For each customer that is represented by sales rep 20 in the
LEVEL1_CUSTOMER table and that also has a balance that does not
exceed the credit limit, increase the customer's credit limit to $8,000.

The only difference between Examples 3 and 4 is that Example 4 uses a compound
condition to identify the rows to be changed. The UPDATE command and the SELECT
command that shows its results appear in Figure 5.5.

FIGURE 5.5

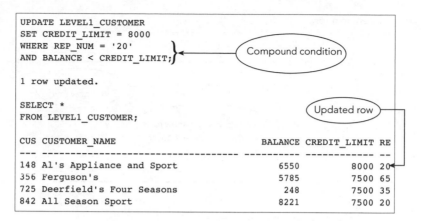

```
UPDATE LEVEL1_CUSTOMER
SET CREDIT_LIMIT = 8000
WHERE REP_NUM = '20'
AND BALANCE < CREDIT_LIMIT;          Compound condition

1 row updated.

SELECT *
FROM LEVEL1_CUSTOMER;                  Updated row

CUS CUSTOMER_NAME                BALANCE CREDIT_LIMIT RE
--- ---------------------------- ------- ------------ --
148 Al's Appliance and Sport        6550         8000 20
356 Ferguson's                      5785         7500 65
725 Deerfield's Four Seasons         248         7500 35
842 All Season Sport                8221         7500 20
```

You also can use the existing value in a column to calculate an update. For example, if you need to increase the credit limit by 10% instead of changing it to a specific value, you can multiply the existing credit limit by 1.10. The following SET clause makes this change:

```
SET CREDIT_LIMIT = CREDIT_LIMIT * 1.10
```

■ Adding New Rows to an Existing Table

In Chapter 2, you used the INSERT command to add data to the tables in the database. You can also use the INSERT command to update table data.

EXAMPLE 5 : Add customer number 895 to the LEVEL1_CUSTOMER table. The name is Peter and Margaret's, the balance is 0, the credit limit is $8,000, and the rep number is 20.

The appropriate INSERT command is shown in Figure 5.6. The SELECT command shown in the figure verifies that the row was successfully added.

FIGURE 5.6   Inserting a row

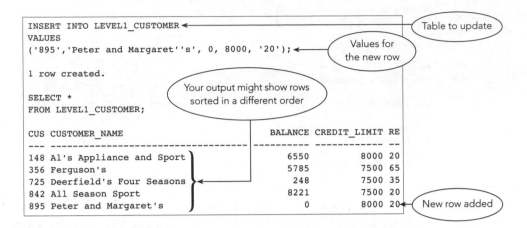

```
INSERT INTO LEVEL1_CUSTOMER                              Table to update
VALUES
('895','Peter and Margaret''s', 0, 8000, '20');         Values for
                                                        the new row
1 row created.

SELECT *                           Your output might show rows
FROM LEVEL1_CUSTOMER;              sorted in a different order

CUS CUSTOMER_NAME                    BALANCE CREDIT_LIMIT RE
--- -------------------------------- ---------- ------------ --
148 Al's Appliance and Sport            6550         8000 20
356 Ferguson's                          5785         7500 65
725 Deerfield's Four Seasons             248         7500 35
842 All Season Sport                    8221         7500 20
895 Peter and Margaret's                   0         8000 20     New row added
```

Note: Your output might be sorted in a different order from what is shown in Figure 5.6. If you need to sort the rows in a specific order, you could use an ORDER BY clause with the desired sort key(s).

■ Commit and Rollback

When you update the data in a table, your updates are only temporary and you can reverse (cancel) them at any time during your current work session. Updates become permanent automatically when you exit from the DBMS. During your current work session, however, you can **commit** (save) your changes immediately by executing the **COMMIT** command.

If you decide that you do not want to save the changes you have made during your current work session, you can **roll back** (reverse) the changes by executing the **ROLLBACK** command. Any updates made since you ran the most recent COMMIT command will be reversed when you run the ROLLBACK command. If you have not run the COMMIT command, executing the ROLLBACK command will reverse all updates made during the current work session. You should note that the ROLLBACK command reverses only changes made to the data; it does not reverse changes made to a table's structure. For example, if you change the length of a character column, you cannot use the ROLLBACK command to return the column length to its original state.

If you do determine that an update was made incorrectly, you can use the ROLLBACK command to return the data to its original state. If, on the other hand, you have verified that the update you made is correct, you can use the COMMIT command to make the update permanent. You do this by typing COMMIT; after running the update. However, you should note that the COMMIT command is permanent; afterwards, running the ROLLBACK command cannot reverse the update.

Figure 5.7 illustrates the execution of the COMMIT command. The SELECT command in the figure verifies that all the previous updates are still reflected in the data. They will still be reflected after any subsequent rollback, as you will see in the next section.

FIGURE 5.7
· · · · · · · · · · · · · · · ·
Performing a
COMMIT

```
COMMIT;

Commit complete.                    All changes
                                    are permanent      COMMIT command
SELECT *
FROM LEVEL1_CUSTOMER;

CUS CUSTOMER_NAME                       BALANCE CREDIT_LIMIT RE
--- ------------------------------- ---------- ------------ --
148 Al's Appliance and Sport              6550         8000 20
356 Ferguson's                            5785         7500 65
725 Deerfield's Four Seasons               248         7500 35
842 All Season Sport                      8221         7500 20
895 Peter and Margaret's                     0         8000 20
```

ACCESS USER : Access does not support the COMMIT or ROLLBACK commands.

MYSQL USER : In order to use the COMMIT and ROLLBACK commands in MySQL, your tables need to have a special type and you also need to change the value for AUTOCOMMIT to 0 (SET AUTOCOMMIT = 0). If you have any questions about using the COMMIT and ROLLBACK commands, check with your instructor.

■ Transactions

A **transaction** is a logical unit of work. A transaction can be viewed as a sequence of steps that accomplishes a single task. It is essential that the entire sequence is completed successfully.

For example, to enter an order, you must add the corresponding order to the ORDERS table, and then add each order line in the order to the ORDER_LINE table. These multiple steps accomplish the "single" task of entering an order. Suppose you have added the order and the first order line, but find you are unable to enter the second order line for some reason; perhaps the part in the order line does not exist. This problem would leave the order in a partially entered state, which is unacceptable. To prevent this problem, you would execute a rollback, thus reversing the insertion of the order and the first order line.

You can use the COMMIT and ROLLBACK commands to support transactions as follows:

- Before beginning the updates for a transaction, commit any previous updates by executing the COMMIT command.
- Complete the updates for the transaction. If any update cannot be completed, execute the ROLLBACK command and discontinue the updates for the current transaction.
- If you can complete all updates successfully, execute the COMMIT command after completing the final update.

■ Deleting Existing Rows from a Table

As you learned in Chapter 2, you use the DELETE command to remove rows from a table. In Example 6, you'll change data and then use the DELETE command to delete a customer from the LEVEL1_CUSTOMER table. In Example 7, you'll execute a rollback to reverse the updates made in Example 6. In this case, the rollback will return the row to its previous state and reinstate the deleted record.

EXAMPLE 6   In the LEVEL1_CUSTOMER table, change the name of customer 356 to Smith Sport, and then delete customer 895.

To delete data from the database, use the DELETE command. The format for the DELETE command is the word DELETE followed by the name of the table from which the row(s) is to be deleted. Next, use a WHERE clause with a condition to select the row(s) to delete. All rows satisfying the condition will be deleted.

The commands shown in Figure 5.8 change the name of customer 356, delete customer 895, and then display the data in the table, verifying the change and the deletion.

FIGURE 5.8

Updating a table

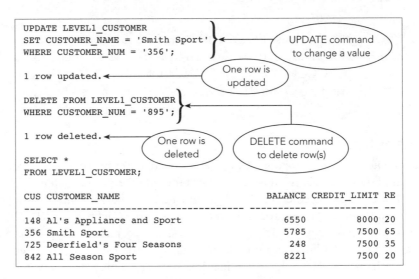

QUESTION   What happens if you run a DELETE command that does not contain a WHERE clause?

ANSWER   Without a condition to specify which row(s) to delete, all rows will be deleted from the table.

■ Executing a Rollback

The following example executes a rollback.

EXAMPLE 7 : Execute a rollback and then display the data in the LEVEL1_CUSTOMER table.

The ROLLBACK command executes a rollback. The command and its results are shown in Figure 5.9. The SELECT command reflects the changes: The name of customer 356 changes back to Ferguson's and the row for customer 895 has been reinstated. All updates made prior to the previous commit are still reflected in the data.

FIGURE 5.9 Performing a rollback

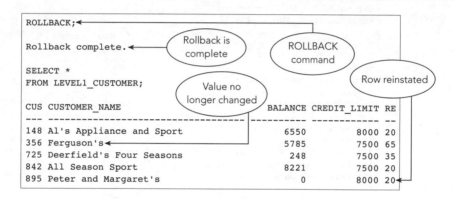

ACCESS USER : If you are using Access to complete these steps, you will not be able to execute the ROLLBACK command. Consequently, your data for the remaining examples in this chapter will differ slightly from the data shown in the figures—customer 356 will be named Smith Sport and customer 895 will not be included.

MYSQL USER : If you are using MySQL to complete these steps, but you are not using the COMMIT and ROLLBACK commands, your data for the remaining examples in this chapter will differ slightly from the data shown in the figures— customer 356 will be named Smith Sport and customer 895 will not be included.

■ Changing a Value in a Column to Null

There are some special issues involved when dealing with nulls. You already have seen how to add a row in which some of the values are null and how to select rows in which a given column is null. You must also be able to change the value in a column in an existing row to null, as shown in Example 8. Remember that in order to make this type of change, the affected column must accept nulls. If you specified NOT NULL for the column when you created the table, then changing a value in a column to null is prohibited.

EXAMPLE 8 : Change the balance of customer 725 in the LEVEL1_CUSTOMER table
to null.

The command for changing the value to null is exactly what it would be for changing any other value. You simply use the value NULL as the replacement value, as shown in Figure 5.10. Notice that the NULL command is *not* enclosed in single quotation marks. If it were, the command would change the balance to the word NULL.

FIGURE 5.10

Setting a column
to null

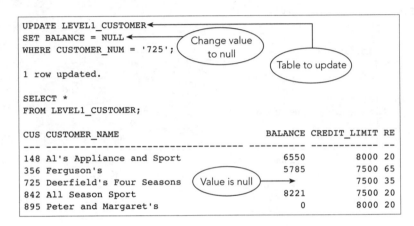

■ Changing a Table's Structure

One of the nicest features of a relational DBMS is the ease with which you can change table structures. You can use the CREATE TABLE command to add new tables, delete tables that are no longer required, add new columns to a table, and change the physical characteristics of existing columns. Next, you will see how to accomplish these changes.

With SQL, you can change a table's structure by using the **ALTER TABLE** command, as illustrated in the following examples.

EXAMPLE 9 : Premiere Products decides to maintain a customer type for each customer
in the database. These types are R for regular customers, D for distributors,
and S for special customers. Add this information in a new column in the
LEVEL1_CUSTOMER table.

To add a new column, use the **ADD** clause of the ALTER TABLE command. The format for the command is the words ALTER TABLE followed by the name of the table to be altered, followed by an appropriate clause. The ADD clause consists of the word ADD followed by the name of the column to be added, followed by the characteristics of the column. Figure 5.11 shows the appropriate ALTER TABLE command for this example.

The LEVEL1_CUSTOMER table now contains a new column named CUSTOMER_TYPE, a CHAR column with a length of 1. Any new rows added to the table must include values for the new column. Effective immediately, all existing rows also contain this new column. The data in any existing row will be changed to reflect the new column the next time the row is updated. However, any time a row is selected for any reason, the system treats the row as though the column is actually present. Thus, to the user, it will seem as though the structure was changed immediately.

For existing rows, some value of CUSTOMER_TYPE must be assigned. The simplest approach (from the point of view of the DBMS, not the user) is to assign the value NULL as a CUSTOMER_TYPE in all existing rows. This process requires the CUSTOMER_TYPE column to accept null values, and some systems actually insist on this. That is, any column added to a table definition *must* accept nulls; the user has no choice. A more flexible approach, and one that is supported by some systems, is to allow the user to specify an initial value. For example, if most customers have type R, you might set the CUSTOMER_TYPE column for existing customers to R, and then later change distributors to type D and special customers to type S. The command to change the structure and set the value in the CUSTOMER_TYPE column to R for all existing records is as follows:

```
ALTER TABLE CUSTOMER
ADD CUSTOMER_TYPE CHAR(1) INIT = 'R';
```

If a system will set new columns only to null, as is the case in Oracle and MySQL, you can still accomplish the previous initialization by following the ALTER TABLE command with an UPDATE command like the one shown in Figure 5.12. The SELECT command in the figure verifies that the value in the CUSTOMER_TYPE column for all rows is R.

FIGURE 5.12 Making the same update for all rows

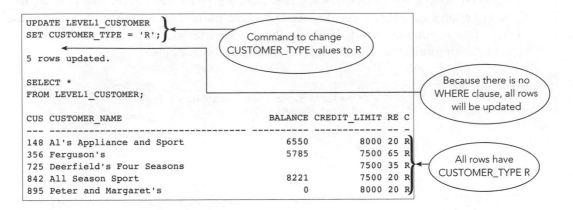

EXAMPLE 10 Two customers in the LEVEL1_CUSTOMER table have a type other than R. Change the types for customers 895 and 148 to S and D, respectively.

The previous example assigned type R to every customer. To change individual types to something other than type R, use the UPDATE command. The appropriate UPDATE commands to make these changes appear in Figure 5.13.

FIGURE 5.13

Updating individual rows

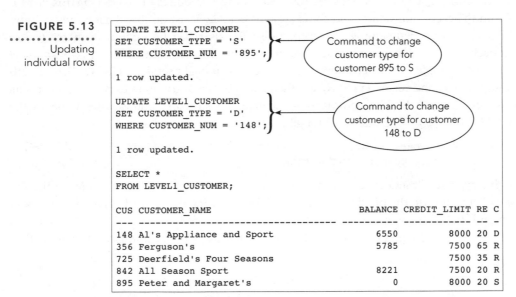

The SELECT command in the figure shows the results of these UPDATE commands. The customer type for customer 895 is S and the type for customer 148 is D. The type for all other customers is R.

Figure 5.14 describes the LEVEL1_CUSTOMER table, including the addition of the CUSTOMER_TYPE column. The table described in Figure 5.14 was produced using the Oracle DESCRIBE command.

FIGURE 5.14   Structure of the LEVEL1_CUSTOMER table

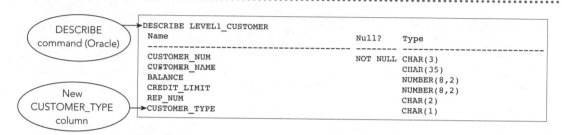

ACCESS
USER

In Access, use the Documenter to show the layout of a table.

MYSQL
USER

In MySQL, use the SHOW COLUMNS command to show the layout of a table. To show the layout of the LEVEL1_CUSTOMER table, for example, the command would be SHOW COLUMNS FROM LEVEL1_CUSTOMER.

EXAMPLE 11

The length of the CUSTOMER_NAME column in the LEVEL1_CUSTOMER table is too short. Increase its length to 50 characters. In addition, change the CREDIT_LIMIT column so that it cannot accept nulls.

You can change the characteristics of existing columns by using the **MODIFY** clause of the ALTER TABLE command. Figure 5.15 shows the ALTER TABLE command that changes the length of the CUSTOMER_NAME column from 35 to 50 characters. The figure also shows the ALTER TABLE command to change the CREDIT_LIMIT column so that it will not accept null values.

FIGURE 5.15

Changing the structure of the LEVEL1_ CUSTOMER table

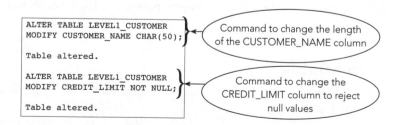

ACCESS
USER

This version of the ALTER TABLE command is not available in Access; to modify the table's structure, make the changes in Design view and save the table.

The DESCRIBE command shown in Figure 5.16 shows the revised structure of the LEVEL1_CUSTOMER table. The length of the CUSTOMER_NAME column is 50 characters, and the CREDIT_LIMIT column will no longer accept nulls.

FIGURE 5.16   Revised structure of the LEVEL1_CUSTOMER table

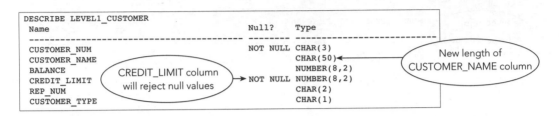

Note: You can also use the MODIFY clause of the ALTER TABLE command to change a column that currently rejects null values so that it will accept null values by using NULL in place of NOT NULL in the ALTER TABLE command.

Note: If there were existing rows in the LEVEL1_CUSTOMER table in which the CREDIT_LIMIT column was already null, the DBMS would reject the modification to the CREDIT_LIMIT column shown in Figure 5.15 and display an error message indicating that this change is not possible. In this case, you must first use an UPDATE command to change all values that are null to some other value. Then you could alter the table's structure as shown in the figure.

■ Making Complex Changes

In some cases, you might need to change a table's structure in ways that are beyond the capabilities of your DBMS. Perhaps you need to eliminate a column, change the column order, or combine data from two tables into one, but your system does not allow these types of changes. For example, some systems, including Oracle, do not allow you to reduce the size of a column or to change a data type. In these situations, you can use the CREATE TABLE command to describe the new table, and then insert values into it using the INSERT command combined with an appropriate SELECT command.

■ Dropping a Table

As you learned in Chapter 2, you can delete a table that is no longer needed by using the DROP TABLE command.

EXAMPLE 12 : Delete the LEVEL1_CUSTOMER table because it is no longer needed in the Premiere Products database.

152

The command to delete the table is shown in Figure 5.17.

FIGURE 5.17
················
Dropping a table

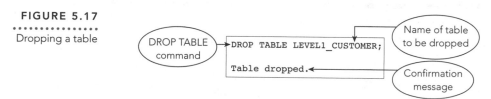

When this command is executed, the LEVEL1_CUSTOMER table and all its data are permanently removed from the database.

In this chapter, you created a new table that contained data from an existing table. You learned how to use the UPDATE command to change the data in a table. You learned to use the INSERT command to add new rows and used the DELETE command to delete existing rows. You learned how to use the COMMIT and ROLLBACK commands to support transactions. You used the ALTER TABLE command to change the structure of a table by adding new columns and changing the characteristics of existing columns. Finally, you learned how to use the DROP TABLE command to delete an entire table and its data. In the next chapter, you will use SQL commands for database administration tasks, including creating views and using the system catalog.

■ SUMMARY

- To create a new table from an existing table, first create the new table by using the CREATE TABLE command. Then use an INSERT command containing a SELECT command to select the desired data to be included from the existing table.

- Use the UPDATE command to change existing data in a table.

- Use the INSERT command to add new rows to a table.

- Use the COMMIT command to make updates permanent; use the ROLLBACK command to reverse any updates that have not been committed.

- Use the DELETE command to delete existing rows from a table.

- To change all values in a column to null, use the SET clause followed by the column name, an equals sign, and the word NULL. To change a specific value in a column to null, use a condition to select the row.

- To add a column to a table, use the ALTER TABLE command with an ADD clause.

- To change the characteristics of a column, use the ALTER TABLE command with a MODIFY clause.

- Use the DROP TABLE command to delete a table and all its data.

■ KEY TERMS

ADD

ALTER TABLE

commit

COMMIT

MODIFY

roll back

ROLLBACK

SET

transaction

UPDATE

■ REVIEW QUESTIONS

1. Which SQL command creates a new table?

2. Which SQL command and clause adds an individual row to a table?

3. How do you add the results of a query to a table?

4. Which SQL command changes data in a table?

5. Which SQL command makes updates permanent?

6. Which SQL command reverses updates? Which updates will be reversed?

7. How do you use the COMMIT and ROLLBACK commands to support transactions?

8. Which SQL command removes rows from a table?

9. What is the format of the SET clause that changes the value of a column to NULL in an UPDATE command?

10. Which SQL command and clause adds a column to an existing table?

11. Which SQL command and clause changes the characteristics of an existing column in a table?

12. Which SQL command deletes a table and all its data?

■EXERCISES (Premiere Products)

Use SQL to make the following changes to the Premiere Products database (see Figure 1.2 in Chapter 1). After each change, execute an appropriate query to show that the change was made correctly. Use the Notes at the end of Chapter 2 to print your output if directed to do so by your instructor.

1. Use the following information to create a new table named NON_APPLIANCE.

Column	Type	Length	Decimal Places	Nulls Allowed?	Description
PART_NUM	Char	4		No	Part number (primary key)
DESCRIPTION	Char	15			Part description
ON_HAND	Decimal	4	0		Number of units on hand
CLASS	Char	2			Item class
PRICE	Decimal	6	2		Unit price

2. Insert into the NON_APPLIANCE table the part number, part description, number of units on hand, item class, and unit price from the PART table for each part that is *not* in item class AP.

3. In the NON_APPLIANCE table, change the description of part number AT94 to "Deluxe Iron."

4. In the NON_APPLIANCE table, increase the price of each item in item class SG by 2%. (*Hint:* Multiply each price by 1.02.)

5. Add the following part to the NON_APPLIANCE table: part number: TL92; description: Trimmer; number of units on hand: 11; class: HW; and price: 29.95.

6. Delete every part in the NON_APPLIANCE table for which the class is SG.

7. In the NON_APPLIANCE table, change the class for part FD21 to null.

8. Add a column named ON_HAND_VALUE to the NON_APPLIANCE table. The on-hand value is a seven-digit number with two decimal places, representing the product of the number of units on hand and the price. Then set all values of ON_HAND_VALUE to ON_HAND * PRICE.

9. In the NON_APPLIANCE table, increase the length of the DESCRIPTION column to 30 characters.

10. Remove the NON_APPLIANCE table from the Premiere Products database.

Use SQL to make the following changes to the Henry Books database (Figures 1.4 through 1.7 in Chapter 1). After each change, execute an appropriate query to show that the change was made correctly. Use the Notes at the end of Chapter 2 to print your output if directed to do so by your instructor.

1. Use the following information to create a new table named MYSTERY.

Column	Type	Length	Decimal Places	Nulls Allowed?	Description
BOOK_CODE	Char	4		No	Book code (primary key)
TITLE	Char	40			Book title
PUBLISHER_CODE	Char	3			Publisher code
PRICE	Decimal	4	2		Unit price

2. Insert into the MYSTERY table the book code, book title, publisher code, and price from the BOOK table for only those books having type MYS.

3. The publisher with code JP has decreased the price of its mystery books by 4%. Update the prices in the MYSTERY table accordingly.

4. Insert a new book into the MYSTERY table. The book code is 9946, the title is *Like Me*, the publisher is MP, and the price is 11.95.

5. Delete the book in the MYSTERY table having book code 9882.

6. The price of the book entitled *The Edge* has been increased to an unknown amount. Change the value in the MYSTERY table to reflect this change.

7. Add to the MYSTERY table a new character column named BEST_SELLER that is one character in length. Then set the default value for all columns to N.

8. Change the BEST_SELLER column in the MYSTERY table to Y for the book entitled *Second Wind*.

9. Change the length of the TITLE column in the MYSTERY table to 50.

10. Change the BEST_SELLER column in the MYSTERY table to reject nulls.

11. Delete the MYSTERY table from the database.

■EXERCISES (Alexamara Marina Group)

Use SQL to make the following changes to the Alexamara Marina Group database (Figures 1.8 through 1.12 in Chapter 1). After each change, execute an appropriate query to show that the change was made correctly. Use the Notes at the end of Chapter 2 to print your output if directed to do so by your instructor.

1. Use the following information to create a new table named LARGE_SLIP. (*Hint:* If you have trouble creating the primary key, see Figure 2.35 in Chapter 2.)

Column	Type	Length	Decimal Places	Nulls Allowed?	Description
MARINA_NUM	Char	4		No	Marina number (primary key)
SLIP_NUM	Char	4		No	Slip number in the marina (primary key)
RENTAL_FEE	Decimal	8	2		Annual rental fee for the slip
BOAT_NAME	Char	50			Name of the boat currently in the slip
OWNER_NUM	Char	4			Number of boat owner renting the slip

2. Insert into the LARGE_SLIP table the marina number, slip number, rental fee, boat name, and owner number for those slips whose length is 40 feet.

3. Alexamara has increased the rental fee of each large slip by $100. Update the rental fees in the LARGE_SLIP table accordingly.

4. After increasing the rental fee of each large slip by $100 (Exercise 3), Alexamara has now decided to decrease the rental fee of any slip whose fee is more than $4,000 by 1%. Update the rental fees in the LARGE_SLIP table accordingly.

5. Insert a new slip into the LARGE_SLIP table. The marina number is 1, the slip number is A4, the rental fee is $3,900.00, the boat name is *Bilsan*, and the owner number is DE62.

6. Delete all slips in the LARGE_SLIP table in which the owner number is TR72.

7. The name of the boat in marina 1 and slip A1 is in the process of being changed to an unknown name. Change the name of this boat in the LARGE_SLIP table to null.

8. Add to the LARGE_SLIP table a new character column named CHARTER that is one character in length. (This column will indicate whether the boat is available to be chartered.) Set the value for the CHARTER column on all rows to N.

9. Change the CHARTER column in the LARGE_SLIP table to Y for the slip containing the boat named *Our Toy*.

10. Change the length of the BOAT_NAME column in the LARGE_SLIP table to 60.

11. Change the RENTAL_FEE column in the LARGE_SLIP table to reject nulls.

12. Delete the LARGE_SLIP table from the database.

CHAPTER 6

Database Administration

OBJECTIVES

- Understand, create, and drop views

- Recognize the benefits of using views

- Grant and revoke users' database privileges

- Understand the purpose, advantages, and disadvantages of using an index

- Create, use, and drop an index

- Understand and obtain information from the system catalog

- Use integrity constraints to control data entry

Introduction

There are some special issues involved in managing a database. This process, often called **database administration**, is especially important when more than one person uses the database. In a business organization, a person or an entire group known as the **database administrator** is charged with managing the database.

In Chapter 5, you learned about one function of the database administrator: changing the structure of a database. In this chapter, you will see how the database administrator can give each user his or her own view of the database. You will use the GRANT and REVOKE commands to assign different database privileges to different users. You will use indexes to improve database performance. You'll learn how SQL stores information about the database structure in a special object called the system catalog and how to access that information. Finally, you'll learn how to specify integrity constraints that establish rules for the data in the database.

■ Views

Most database management systems support the creation of views. A **view** is an application program's or an individual user's picture of the database. The existing, permanent tables in a relational database are called **base tables**. A view is a derived table because the data in it is derived from the base table. To the user, a view appears to be an actual table, but it is not. In many cases, a user can examine table data using a view. Because a view usually includes less information than the full database, its use can represent a great simplification. Views also provide a measure of security, because omitting sensitive tables or columns from a view will render them unavailable to anyone accessing the database through the view.

To help you understand the idea of a view, suppose that Juan is interested in the part number, part description, units on hand, and unit price of parts in item class HW. He is not interested in any other columns in the PART table, nor is he interested in any rows that correspond to parts in other item classes. Viewing this data would be simpler for Juan if the other rows and columns were not even present. Although you cannot change the structure of the PART table and omit some of its rows just for Juan, you can do the next best thing. You can provide him with a view that consists of only the rows and columns that he needs to access.

A view is defined by creating a **defining query**, which indicates the rows and columns to include in the view. The SQL command (or the defining query) to create the view for Juan is illustrated in Example 1.

M**YSQL** : MySQL does not currently support views. Support for views is planned for
USER : version 5.0. If you are using MySQL, you will not be able to complete the
: steps in this section, but it is very important that you read it so that you
: understand this valuable concept.

E**XAMPLE 1** : Define a view named HOUSEWARES that consists of the part number, part
: description, units on hand, and unit price of each part in item class HW.

To create a view definition, use the **CREATE VIEW** command, which includes the words CREATE VIEW, followed by the name of the view, the word AS, and then a query. The CREATE VIEW command shown in Figure 6.1 creates a view of the PART table that contains only the specified columns.

FIGURE 6.1   Creating the HOUSEWARES view

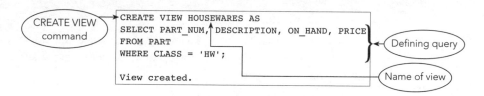

A**CCESS** : In Access, you do not use the CREATE VIEW command to create a view.
USER : Instead, you create the defining query and then save the query using the
: view's name (for example, HOUSEWARES).

Given the current data in the Premiere Products database, the HOUSEWARES view contains the data shown in Figure 6.2.

FIGURE 6.2
HOUSEWARES
view

HOUSEWARES

PART_NUM	DESCRIPTION	ON_HAND	PRICE
AT94	Iron	50	$24.95
DL71	Cordless Drill	21	$129.95
FD21	Stand Mixer	22	$159.95

The data does not actually exist in this form, however, nor will it *ever* exist in this form. It is tempting to think that when this view is used, the query is executed and produces some sort of temporary table, named HOUSEWARES, which Juan then could access, but this is *not* what actually happens. Instead, the query acts as a sort of "window" into the database, as shown in Figure 6.3. As far as Juan is concerned, the entire database is just the darker shaded portion of the PART table. Juan can see any change that affects the darker portion of the PART table, but he is totally unaware of any other changes that are made in the database.

FIGURE 6.3   Juan's view of the PART table

PART

PART_NUM	DESCRIPTION	ON_HAND	CLASS	WAREHOUSE	PRICE
AT94	Iron	50	HW	3	$24.95
BV06	Home Gym	45	SG	2	$794.95
CD52	Microwave Oven	32	AP	1	$165.00
DL71	Cordless Drill	21	HW	3	$129.95
DR93	Gas Range	8	AP	2	$495.00
DW11	Washer	12	AP	3	$399.99
FD21	Stand Mixer	22	HW	3	$159.95
KL62	Dryer	12	AP	1	$349.95
KT03	Dishwasher	8	AP	3	$595.00
KV29	Treadmill	9	SG	2	$1,390.00

When you create a query that involves a view, SQL changes the query to one that selects data from the table(s) in the database that created the view. For example, suppose Juan creates the query shown in Figure 6.4.

FIGURE 6.4   Using the HOUSEWARES view

The DBMS does not execute the query in this form. Instead, it merges the query Juan entered with the query that defines the view to form the query that is actually executed. When the DBMS merges the query that creates the view with the query to select rows where the ON_HAND value is more than 10, the query that SQL actually executes is:

```
SELECT PART_NUM, DESCRIPTION, ON_HAND, PRICE
FROM PART
WHERE CLASS = 'HW'
AND ON_HAND > 10;
```

In the query that the DBMS executes, the FROM clause lists the PART table rather than the HOUSEWARES view, the SELECT clause lists columns from the PART table instead of * to select all columns from the HOUSEWARES view, and the WHERE clause contains a compound condition to select only those parts in the HW class (as Juan sees in the HOUSEWARES view) and only those parts with ON_HAND values of more than 10. This new query is the one that the DBMS actually executes.

Juan, however, is unaware that this kind of activity is taking place. To Juan, it seems that he is really using a table named HOUSEWARES. One advantage of this approach is that because the HOUSEWARES view never exists in its own right, any update to the PART table is *immediately* available in the HOUSEWARES view. If the HOUSEWARES view were really a table, this immediate update would not be possible.

You also can assign column names that are different from those in the base table, as illustrated in the next example.

EXAMPLE 2 : Define a view named HSEWRES that consists of the part number, part description, units on hand, and unit price of all parts in item class HW. In this view, change the names of the PART_NUM, DESCRIPTION, ON_HAND, and PRICE columns to NUM, DSC, OH, and PRCE, respectively.

When renaming columns, you include the new column names in parentheses following the name of the view, as shown in Figure 6.5. In this case, anyone accessing the HSEWRES view will refer to PART_NUM as NUM, to DESCRIPTION as DSC, to ON_HAND as OH, and to PRICE as PRCE. If you select all columns from the HSEWRES view, the output will display the new column names, as shown in Figure 6.5.

FIGURE 6.5   Renaming columns when creating a view

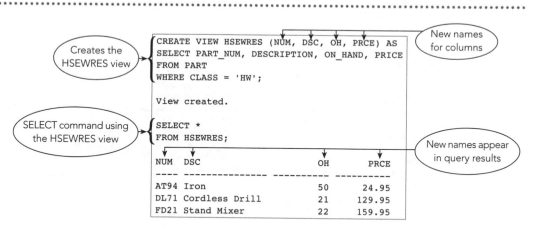

ACCESS : To change column names in Access, use AS clauses in the query design (for USER : example, SELECT PART_NUM AS NUM, DESCRIPTION AS DSC, and so on).

The HSEWRES view is an example of a **row-and-column subset view** because it consists of a subset of the rows and columns in some base table—in this case, in the PART table. Because the defining query can be any valid SQL query, a view could also join two or more tables or involve statistics. The next example illustrates a view that joins two tables.

: Define a view named REP_CUST consisting of the sales rep number
    (named RNUM), sales rep last name (named RLAST), sales rep first name
    (named RFIRST), customer number (named CNUM), and customer name
    (named CNAME) for all sales reps and matching customers in the REP and
    CUSTOMER tables.

The command to create this view appears in Figure 6.6.

FIGURE 6.6 Creating the REP_CUST view

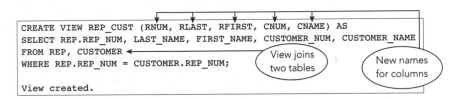

```
CREATE VIEW REP_CUST (RNUM, RLAST, RFIRST, CNUM, CNAME) AS
SELECT REP.REP_NUM, LAST_NAME, FIRST_NAME, CUSTOMER_NUM, CUSTOMER_NAME
FROM REP, CUSTOMER                          View joins          New names
WHERE REP.REP_NUM = CUSTOMER.REP_NUM;       two tables          for columns

View created.
```

Given the current data in the Premiere Products database, the REP_CUST view contains the data shown in Figure 6.7.

FIGURE 6.7 Using the REP_CUST view

```
SELECT *
FROM REP_CUST;

RN RLAST            RFIRST           CNU CNAME
-- --------------   --------------   --- ----------------------------
20 Kaiser           Valerie          148 Al's Appliance and Sport
35 Hull             Richard          282 Brookings Direct
65 Perez            Juan             356 Ferguson's
35 Hull             Richard          408 The Everything Shop
65 Perez            Juan             462 Bargains Galore
20 Kaiser           Valerie          524 Kline's
65 Perez            Juan             608 Johnson's Department Store
35 Hull             Richard          687 Lee's Sport and Appliance
35 Hull             Richard          725 Deerfield's Four Seasons
20 Kaiser           Valerie          842 All Season
```

A view can also involve statistics.

EXAMPLE 4 : Define a view named CRED_CUST that consists of each credit limit
    (CREDIT_LIMIT) and the number of customers having this credit limit
    (NUM_CUSTOMERS).

The command shown in Figure 6.8 creates this view; the SELECT command displays
the current data in the Premiere Products database for this view.

164

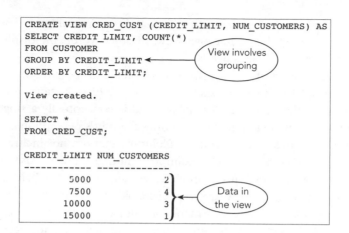

FIGURE 6.8
· · · · · · · · · · · · · · · ·
Creating the
CRED_CUST
view

The use of views provides several benefits. First, views provide data independence. If the database structure changes (by adding columns or changing the way objects are related, for example) in such a way that the view can still be derived from existing data, the user can still access and use the same view. If adding extra columns to tables in the database is the only change, and these columns are not required by the view's user, the defining query might not even need to be changed in order for the user to continue using the view. If relationships are changed, the defining query might be different, but because users aren't aware of the defining query, they are unaware of this difference. Users continue accessing the database through the same view, as though nothing has changed. For example, suppose that customers are assigned to territories, that each territory is assigned to a single sales rep, that a sales rep can have more than one territory, and that a customer is represented by the sales rep who covers the customer's assigned territory. To implement these changes, you might choose to restructure the database as follows:

```
REP(REP_NUM, LAST_NAME, FIRST_NAME, STREET, CITY, STATE, ZIP,
     COMMISSION, RATE)
TERRITORY(TERRITORY_NUM, DESCRIPTION, REP_NUM)
CUSTOMER(CUSTOMER_NUM, CUSTOMER_NAME, STREET, CITY, STATE, ZIP, BALANCE,
     CREDIT_LIMIT, TERRITORY_NUM)
```

Assuming that the REP_CUST view shown earlier is still required, the defining query could be changed as follows:

```
CREATE VIEW REP_CUST (RNUM, RLAST, RFIRST, CNUM, CNAME) AS
SELECT REP.REP_NUM, REP.LAST_NAME, REP.FIRST_NAME, CUSTOMER_NUM, CUSTOMER_NAME
FROM REP, TERRITORY, CUSTOMER
WHERE REP.REP_NUM = TERRITORY.REP_NUM
AND TERRITORY.TERRITORY_NUM = CUSTOMER.TERRITORY_NUM;
```

The user of this view can still retrieve the number and name of a sales rep together with the number and name of each customer the sales rep represents. The user will be unaware, however, of the new structure in the database.

The second benefit of using views is that different users can view the same data in different ways through his or her own view. In other words, the display of data can be customized to meet each user's needs.

The final benefit of using views is that a view can contain only those columns required by a given user. This practice has two advantages. First, because the view usually contains fewer columns than the overall database and is conceptually a single table, rather than a collection of tables, a view greatly simplifies the user's perception of the database. Second, views provide a measure of security. Columns that are not included in the view are not accessible to the view's user. For example, omitting the BALANCE column from a view will ensure that the view's user cannot access any customer's balance. Likewise, rows that are not included in the view are not accessible. A user of the HOUSEWARES view, for example, cannot obtain any information about parts in the AP or SG classes.

These benefits hold true only when views are used for retrieval purposes. When updating the database, the issues involved in updating data through a view depend on the type of view, as you will see next.

Row-and-Column Subsets

Consider the row-and-column subset view for the HOUSEWARES view. There are columns in the underlying base table (PART) that are not present in the view. Thus, if you attempt to add a row with the data 'BB99','PAN',50,14.95, the DBMS must determine how to enter the data in those columns from the PART table that are not included in the HOUSEWARES view (CLASS and WAREHOUSE). In this case, it is clear what data to enter in the CLASS column. According to the view definition, all rows are item class HW, but it is not clear what data to enter in the WAREHOUSE column. The only possibility would be NULL. Thus, provided that every column not included in a view can accept nulls, you can add new rows using the INSERT command. There is another problem, however. Suppose the user attempts to add a row to the HOUSEWARES view containing the data 'BV06','Waffle Maker',5,29.95. Because part number BV06 already exists in the PART table, the system *must* reject this attempt. Because this part is not in item class HW (and therefore is not in the HOUSEWARES view), this rejection certainly will seem strange to the user, because there is no such part in the user's view.

On the other hand, updates or deletions cause no particular problem in this view. If the description of part number FD21 changes from Stand Mixer to Pan, this change is made in the PART table. If part number DL71 is deleted, this deletion occurs in the PART table. One surprising change could take place, however. Suppose that the CLASS column is included in the HOUSEWARES view and a user changes the class of part number AT94 from HW to AP. Because this item would no longer satisfy the criterion for being included in the HOUSEWARES view, part number AT94 would disappear from the user's view!

Although there are problems to overcome, it seems possible to update the database through the HOUSEWARES view. This does not imply that *any* row-and-column subset view is updatable, however. Consider the view and data shown in Figure 6.9. (The DISTINCT operator omits duplicate rows from the view.)

FIGURE 6.9
· · · · · · · · · · · · · · · ·
Creating the
REP_CRED view

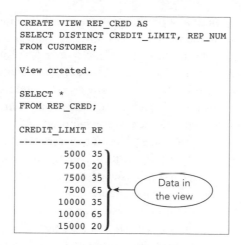

```
CREATE VIEW REP_CRED AS
SELECT DISTINCT CREDIT_LIMIT, REP_NUM
FROM CUSTOMER;

View created.

SELECT *
FROM REP_CRED;

CREDIT_LIMIT RE
------------ --
        5000 35
        7500 20
        7500 35
        7500 65
       10000 35
       10000 65
       15000 20
```

Data in the view

How would you add the row 15000,'35' to this view? In the underlying base table (CUSTOMER), at least one customer must be added whose credit limit is $15,000 and whose sales rep number is 35, but which customer is it? You can't leave the other columns null in this case, because one of them is CUSTOMER_NUM, which is the base table's primary key. What would it mean to change the row 5000,'35' to 15000,'35'? Would it mean changing the credit limit to $15,000 for each customer represented by sales rep number 35 that currently has a credit limit of $5,000? Would it mean changing the credit limit of one of these customers and deleting the rest? What would it mean deleting the row 5000,'35'? Would it mean deleting all customers with credit limits of $5,000 and represented by sales rep number 35, or would it mean assigning these customers a different sales rep or a different credit limit?

Why does the REP_CRED view involve a number of serious problems that are not present in the HOUSEWARES view? The basic reason is that the HOUSEWARES view includes, as one of its columns, the primary key of the underlying base table, but the REP_CRED view does not. A row-and-column subset view that contains the primary key of the underlying base table is updatable (subject, of course, to some of the concerns already discussed).

Joins

In general, views that involve joins of base tables can cause problems at update. Consider the relatively simple REP_CUST view, for example, described earlier (see Figures 6.6 and 6.7). The fact that some columns in the underlying base tables are not seen in this view presents some of the same problems discussed earlier. Even assuming that these problems can be overcome through the use of nulls, there are more serious problems when attempting to update the database through this view. On the surface, changing the row '35','Hull','Richard','282','Brookings Direct' to '35','Baldwin','Sara','282','Brookings Direct', might not appear to pose any problems other than some inconsistency in the data. (In the new version of the row, the name of sales rep 35 is Sara Baldwin; whereas in the fourth row in the table, the name of sales rep 35, *the same sales rep*, is Richard Hull.)

The problem is actually more serious than that—making this change is not possible. The name of a sales rep is stored only once in the underlying REP table. Changing the name of sales rep 35 from Richard Hull to Sara Baldwin in this one row of the view causes the change to be made to the single row for sales rep 35 in the REP table. Because the view simply displays data from the base tables, for each row in which the sales rep number is 35, the sales rep name is now Sara Baldwin. In other words, it appears that the same change has been made in the other rows. In this case, this change ensures consistency in the data. In general, however, the unexpected changes caused by an update are not desirable.

Before concluding the topic of views that involve joins, you should note that all joins do not create the preceding problem. If two base tables have the same primary key and the primary key is used as the join column, updating the database is not a problem. For example, suppose that the actual database contains two tables (REP_DEMO and REP_FIN) instead of one table (REP). Figure 6.10 shows the data in these hypothetical tables.

FIGURE 6.10   REP_DEMO and REP_FIN tables

REP_DEMO

REP_NUM	LAST_NAME	FIRST_NAME	STREET	CITY	STATE	ZIP
20	Kaiser	Valerie	624 Randall	Grove	FL	33321
35	Hull	Richard	532 Jackson	Sheldon	FL	33553
65	Perez	Juan	1626 Taylor	Fillmore	FL	33336

REP_FIN

REP_NUM	COMMISSION	RATE
20	$20,542.50	0.05
35	$39,216.00	0.07
65	$23,487.00	0.05

In this case, what was a single table in the original Premiere Products design has been divided into two separate tables. Users who need to see the rep data in a single table can use a view that joins these two tables using the REP_NUM column. The view definition appears in Figure 6.11.

FIGURE 6.11 Creating the SALES_REP view

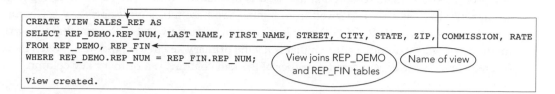

```
CREATE VIEW SALES_REP AS
SELECT REP_DEMO.REP_NUM, LAST_NAME, FIRST_NAME, STREET, CITY, STATE, ZIP, COMMISSION, RATE
FROM REP_DEMO, REP_FIN
WHERE REP_DEMO.REP_NUM = REP_FIN.REP_NUM;

View created.
```

View joins REP_DEMO and REP_FIN tables

Name of view

The SALES_REP view appears in Figure 6.12.

FIGURE 6.12 Using the SALES_REP view

```
SELECT *
FROM SALES_REP;

RE LAST_NAME        FIRST_NAME       STREET           CITY              ST ZIP    COMMISSION     RATE
-- ---------------- ---------------- ---------------- ---------------- -- ----- ---------- ---------
20 Kaiser           Valerie          624 Randall      Grove             FL 33321    20542.5       .05
35 Hull             Richard          532 Jackson      Sheldon           FL 33553      39216       .07
65 Perez            Juan             1626 Taylor      Fillmore          FL 33336      23487       .05
```

It's easy to update the SALES_REP view. To add a row, simply add a row to each underlying base table. To change data in a row, change the appropriate base table. To delete a row from the view, delete the corresponding rows from both underlying base tables.

QUESTION How would you add the row '10','Peters','Jean','14 Brink','Holt','FL','46223',107.50,.05 to the SALES_REP view?

ANSWER Add the row '10','Peters','Jean','14 Brink','Holt','FL','46223' to the REP_DEMO table, and then add the row '10',107.50,0.05 to the REP_FIN table.

QUESTION How would you change the name of sales rep 20 to Valerie Lewis?

ANSWER Use an update to change the name in the REP_DEMO table.

QUESTION How would you change Valerie's commission rate to 0.06?

ANSWER Use an update to change the rate in the REP_FIN table.

QUESTION How would you delete sales rep 35 from the REP table?

ANSWER Delete sales rep 35 from *both* the REP_DEMO *and* REP_FIN tables.

Updates (additions, changes, or deletions) to the SALES_REP view would not cause any problems. The main reason that this view is updatable—and other views involving joins are not—is that this view is derived from joining two base tables *on the primary key of each table*. In contrast, the REP_CUST view is derived by joining two tables by matching the primary key of one table with a column that is *not* the primary key in the other table. Even more severe problems are encountered if neither of the join columns is a primary key column.

Statistics

A view that involves statistics calculated from one or more base tables is the most troublesome view of all. Consider the CRED_CUST view, for example (see Figure 6.8). How would you add the row 9000,3 to indicate that there are three customers that have credit limits of $9,000 each? Likewise, changing the row 5000,2 to 5000,5 means you are adding three new customers with credit limits of $5,000 each, for a total of five customers. Clearly these are impossible tasks; you can't add rows to a view that includes calculations.

Dropping a View

When a view is no longer needed, you can remove it using the **DROP VIEW** command.

EXAMPLE 5 : The HSEWRES view is no longer necessary, so delete it.

The DROP VIEW command to delete this view is shown in Figure 6.13. The DROP VIEW command removes only a view definition; the tables and data on which the view is based still exist.

FIGURE 6.13
··················
Dropping a view

```
DROP VIEW HSEWRES;

View dropped.
```

ACCESS : Rather than using the DROP VIEW command, you simply delete the query
USER : you saved when you created the view.

■ Security

Security is the prevention of unauthorized access to a database. Within an organization, the database administrator determines the types of access various users can have to the database. Some users may be able to retrieve and update anything in the database. Other users may be able to retrieve any data from the database but not make any changes to it. Still other users may be able to access only a portion of the database. For example, Bill may be able to retrieve and update customer data, but is prevented from accessing data about sales reps, orders, order lines, or parts. Valerie may be able to retrieve part data and nothing else. Sam may be able to retrieve and update data on parts in the HW class, but cannot retrieve data in any other classes.

After the database administrator has determined the access different users of the database will have, these access rules are enforced by whatever security mechanism the DBMS provides. In SQL systems, there are two security mechanisms. You already have seen that views furnish a certain amount of security; if users are accessing the database through a view, they cannot access any data that is not included in the view. The main mechanism for providing access to a database, however, is the **GRANT** command.

The basic idea is that the database administrator can grant different types of privileges to users and then revoke them later, if necessary. These privileges include the right to select, insert, update, and delete table data. You can grant and revoke user privileges using the GRANT and REVOKE commands. The following examples illustrate various uses of the GRANT command when the named users already exist in the database.

Note: You should not attempt to execute the commands in this section unless your instructor specifically directs you to do so.

EXAMPLE 6 : User Johnson must be able to retrieve data from the REP table.

The following GRANT command permits a user named Johnson to execute SELECT commands for the REP table:

```
GRANT SELECT ON REP TO Johnson;
```

EXAMPLE 7 : Users Smith and Brown must be able to add new parts to the PART table.

The following GRANT command permits two users named Smith and Brown to execute INSERT commands for the PART table. Notice that a comma separates the users' names.

```
GRANT INSERT ON PART TO Smith, Brown;
```

EXAMPLE 8 : User Anderson must be able to change the name and street address of customers.

The following GRANT command permits a user named Anderson to execute UPDATE commands involving the CUSTOMER_NAME and STREET columns in the CUSTOMER table. Notice that the SQL command includes the table name followed by the column name(s) to update in parentheses.

```
GRANT UPDATE ON CUSTOMER (CUSTOMER_NAME, STREET) TO Anderson;
```

EXAMPLE 9 : User Thompson must be able to delete order lines.

The following GRANT command permits a user named Thompson to execute a DELETE command for the ORDER_LINE table:

```
GRANT DELETE ON ORDER_LINE TO Thompson;
```

EXAMPLE 10 : Every user must be able to retrieve part numbers, part descriptions, and item classes.

The GRANT command to indicate that all users may retrieve data using a SELECT command includes the word PUBLIC, as follows:

```
GRANT SELECT ON PART (PART_NUM, DESCRIPTION, CLASS) TO PUBLIC;
```

EXAMPLE 11 : User Roberts must be able to create an index on the REP table.

You will learn about indexes and their uses in the next section. The following GRANT command permits a user named Roberts to create an index on the REP table:

```
GRANT INDEX ON REP TO Roberts;
```

EXAMPLE 12 : User Thomas must be able to change the structure of the CUSTOMER table.

The following GRANT command permits a user named Thomas to execute ALTER commands for the CUSTOMER table to change the table's structure.

```
GRANT ALTER ON CUSTOMER TO Thomas;
```

EXAMPLE 13 : User Wilson must have all privileges for the REP, CUSTOMER, and ORDERS tables.

The GRANT command to indicate that a user has all privileges includes the ALL privilege.

```
GRANT ALL ON REP, CUSTOMER, ORDERS TO Wilson;
```

The privileges that can be granted are SELECT to retrieve data, UPDATE to change data, DELETE to delete data, INSERT to add new data, INDEX to create an index, and ALTER to change the table structure. The database administrator usually assigns privileges. Normally, when the database administrator grants a particular privilege to a user, the user cannot pass that privilege along to other users. If the user needs to be able to pass the privilege to other users, the GRANT command must include the **WITH GRANT OPTION** clause. This clause grants the indicated privilege to the user and also permits the user to grant the same privileges (or a subset of them) to other users.

The database administrator uses the **REVOKE** command to revoke privileges from users. The format of the REVOKE command is essentially the same as that of the GRANT command, but with two differences: the word GRANT is replaced by the word REVOKE and the word TO is replaced by the word FROM. In addition, the clause WITH GRANT OPTION obviously is not meaningful as part of a REVOKE command. Incidentally, the revoke will cascade so that if Johnson is granted privileges WITH GRANT OPTION and then Johnson grants these same privileges to Smith, revoking the privileges from Johnson revokes Smith's privileges at the same time. Example 14 illustrates the use of the REVOKE command.

EXAMPLE 14 : User Johnson is no longer allowed to retrieve data from the REP table.

The following REVOKE command revokes the SELECT privilege from the user named Johnson:

```
REVOKE SELECT ON REP FROM Johnson;
```

The GRANT and REVOKE commands also can be applied to views so that access is restricted to only certain rows within tables.

EXAMPLE 15 : Allow sales rep 20 (Valerie Kaiser) to access any data concerning the customers she represents, but restrict her from accessing data about any other customer.

You cannot include a WHERE clause in a GRANT command. You must first create a view containing exactly the data the user can access and then grant the user the appropriate privileges for the view. The SQL command to do so are follows:

```
CREATE VIEW REP20CST AS
SELECT *
FROM CUSTOMER
WHERE REP_NUM = '20'
GRANT SELECT ON REP20CST TO Kaiser;
```

■ Indexes

When you query a database, you're usually searching for a row (or collection of rows) that satisfies some condition. Examining every row in a table to find the ones you need often takes too much time to be practical, especially in tables with thousands of records. Fortunately, you can create and use an index to speed up the searching process significantly. An index in a DBMS is similar to an index in a book. If you want to find a discussion of a given topic in a book, you could scan the entire book from start to finish, looking for references to the topic you need. More than likely, however, you wouldn't have to resort to this technique. If the book has a good index, you could use it to identify the pages on which your topic is discussed.

Within relational model systems on both mainframes and personal computers, the main mechanism for increasing the efficiency with which data is retrieved from the database is the **index**. Conceptually, these indexes are very much like the index in a book. Consider Figure 6.14, for example, which shows the CUSTOMER table for Premiere Products together with one extra column named ROW_NUMBER. This extra column gives the location of the row in the table (customer 148 is the first row in the table and is on row 1, customer 282 is on row 2, and so on). These row numbers are automatically assigned and used by the DBMS, not by the users, and that is why you do not normally see them.

FIGURE 6.14 CUSTOMER table with row numbers

CUSTOMER

ROW_ NUMBER	CUSTOMER_ NUM	CUSTOMER_ NAME	STREET	CITY	STATE	ZIP	BALANCE	CREDIT_ LIMIT	REP_ NUM
1	148	Al's Appliance and Sport	2837 Greenway	Fillmore	FL	33336	$6,550.00	$7,500.00	20
2	282	Brookings Direct	3827 Devon	Grove	FL	33321	$431.50	$10,000.00	35
3	356	Ferguson's	382 Wildwood	Northfield	FL	33146	$5,785.00	$7,500.00	65
4	408	The Everything Shop	1828 Raven	Crystal	FL	33503	$5,285.25	$5,000.00	35
5	462	Bargains Galore	3829 Central	Grove	FL	33321	$3,412.00	$10,000.00	65
6	524	Kline's	838 Ridgeland	Fillmore	FL	33336	$12,762.00	$15,000.00	20
7	608	Johnson's Department Store	372 Oxford	Sheldon	FL	33553	$2,106.00	$10,000.00	65
8	687	Lee's Sport and Appliance	282 Evergreen	Altonville	FL	32543	$2,851.00	$5,000.00	35
9	725	Deerfield's Four Seasons	282 Columbia	Sheldon	FL	33553	$248.00	$7,500.00	35
10	842	All Season	28 Lakeview	Grove	FL	33321	$8,221.00	$7,500.00	20

To access a customer's row using its customer number, you might create and use an index, as shown in Figure 6.15. The index has two columns. The first column contains a customer number and the second column contains the number of the row on which the customer number is found. To find a customer, you look up the customer's number in the first column in the index. The value in the second column indicates which row to retrieve from the CUSTOMER table; then the row for the desired customer is retrieved.

FIGURE 6.15
....
Index for
CUSTOMER
table on
CUSTOMER_
NUM column

CUSTOMER_NUM INDEX

CUSTOMER_NUM	ROW_NUMBER
148	1
282	2
356	3
408	4
462	5
524	6
608	7
687	8
725	9
842	10

Because customer numbers are unique, there is only a single row number in this index. This is not always the case, however. Suppose that you wanted to access all customers with a specific credit limit or all customers that are represented by a specific sales rep. You might choose to create and use an index on the CREDIT_LIMIT column and an index on the REP_NUM column, as shown in Figure 6.16. In the CREDIT_LIMIT index, the first column contains a credit limit and the second column contains the numbers of *all* rows on which that credit limit is found. The REP_NUM index is similar, except that the first column contains a sales rep number.

FIGURE 6.16
....
Indexes for
CUSTOMER
table on
CREDIT_LIMIT
and REP_NUM
columns

CREDIT_LIMIT INDEX

CREDIT_LIMIT	ROW_NUMBER
$5,000.00	4, 8
$7,500.00	1, 3, 9, 10
$10,000.00	2, 5, 7
$15,000.00	6

REP_NUM INDEX

REP_NUM	ROW_NUMBER
20	1, 6, 10
35	2, 4, 8, 9
65	3, 5, 7

QUESTION How would you use the index shown in Figure 6.16 to find every customer with a $10,000 credit limit?

ANSWER Look up $10,000 in the CREDIT_LIMIT index to find a collection of row numbers (2, 5, and 7). Use these row numbers to find the corresponding rows in the CUSTOMER table (Brookings Direct, Bargains Galore, and Johnson's Department Store).

QUESTION   How would you use the index shown in Figure 6.16 to find every cus-
           tomer represented by sales rep 35?

ANSWER   Look up 35 in the REP_NUM index to find a collection of row numbers (2,
           4, 8, and 9). Use these row numbers to find the corresponding rows in the
           CUSTOMER table (Brookings Direct, The Everything Shop, Lee's Sport
           and Appliance, and Deerfield's Four Seasons).

The actual structure of an index is more complicated than what is shown in the figures. Fortunately, you don't have to be concerned with the details of manipulating and
using indexes because the DBMS manages them for you—your only job is to determine
the columns on which to build the indexes. Typically, you can create and maintain an index
for any column or combination of columns in any table. After creating an index, the
DBMS uses it to speed up data retrieval.

As you would expect, the use of any index has advantages and disadvantages. An
important advantage was already mentioned: An index makes certain types of retrieval
more efficient.

There are two disadvantages. First, an index occupies disk space. Using this space for
an index, however, is technically unnecessary because any retrieval that you can make
using an index can also be made without the index; the index just speeds up the retrieval.
The second disadvantage is that the DBMS must update the index whenever corresponding data in the database is updated. Without the index, the DBMS would not need to make
these updates. The main question that you must ask when considering whether to create a
given index is this: Do the benefits derived during retrieval outweigh the additional storage
required and the extra processing involved in update operations? In a very large database,
you might find that indexes are essential to decrease the time required to retrieve records.
However, in a small database, an index might not provide any significant benefits.

You can add and drop indexes as necessary. You can create an index after the database
is built; it doesn't need to be created at the same time as the database. Likewise, if it
appears that an existing index is unnecessary, you can drop it.

Creating an Index

Suppose that some users at Premiere Products need to display customer records ordered
by balance. Other users need to access a customer's name using the customer's number. In
addition, some users need to produce a report in which customer records are listed by
credit limit in descending order. Within the group of customers having the same credit
limit, the customer records must be ordered by name.

Each of the previous requirements is carried out more efficiently when you create the
appropriate index. The command used to create an index is **CREATE INDEX,** as illustrated
in Example 16.

EXAMPLE 16 : Create an index named BALIND on the BALANCE column in the CUSTOMER table. Create an index named REPNAME on the combination of the LAST_NAME and FIRST_NAME columns in the REP table. Create an index named CREDNAME on the combination of the CREDIT_LIMIT and CUSTOMER_NAME columns in the CUSTOMER table, with the credit limits listed in descending order.

The appropriate CREATE INDEX commands to create these indexes appear in Figure 6.17. Each command lists the name of the index and the table name on which the index is to be created. The column name(s) are listed in parentheses. If any column is to be included in descending order, the column name is followed by the word DESC.

FIGURE 6.17
....
Creating indexes

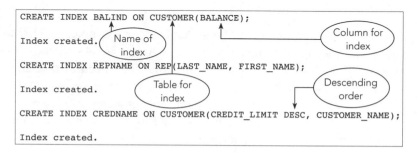

If customers are listed using the CREDNAME index, the records will appear in order by descending credit limit. Within any credit limit, the customers will be listed alphabetically by name.

ACCESS : Access supports the creation of indexes in both SQL view and Design
USER : view. You can use an SQL command with the CREATE INDEX command in SQL view. In addition, if you open the table containing the column(s) on which you want to create an index in Design view, you can click the Indexes button or use the Indexed property to create the index.

Dropping an Index

The command used to drop (delete) an index is **DROP INDEX**, which consists of the words DROP INDEX followed by the name of the index to drop. To delete the CREDNAME index, for example, the command would be:

```
DROP INDEX CREDNAME;
```

The DROP INDEX command permanently deletes the index. CREDNAME was the index the DBMS used when listing customer records in descending credit limit order and then by customer name within credit limit. The DBMS still can list customers in this order; however, it cannot do so as efficiently without the index.

Unique Indexes

When you specify a table's primary key, the DBMS automatically ensures that the values entered in the primary key column(s) are unique. For example, the DBMS would reject an attempt to add a second customer whose number is 148 in the CUSTOMER table because customer 148 already exists. Thus, you don't need to take any special action to make sure that values in the primary key column are unique; the DBMS does it for you.

Occasionally, a non-primary key column might store unique values. For example, in the REP table, the primary key is REP_NUM. If the REP table also contains a column for Social Security numbers, the values in this column also must be unique. Because the Social Security number column is not the table's primary key, however, you need to take special action in order for the DBMS to ensure that there are no duplicate values in this column.

To ensure the uniqueness of values in a non-primary key column, you can create a **unique index** by using the **CREATE UNIQUE INDEX** command. To create a unique index named SSN on the SOC_SEC_NUM column of the REP table, for example, the command would be:

```
CREATE UNIQUE INDEX SSN ON REP(SOC_SEC_NUM);
```

The unique index has all the properties of indexes already discussed, along with one additional property: The DBMS will reject any update that would cause a duplicate value in the SOC_SEC_NUM column. In this case, the DBMS will reject the addition of a rep whose Social Security number is the same as that of another rep already in the database.

■ System Catalog

Information about tables in the database is kept in the **system catalog** (**catalog**) or the **data dictionary**. This section describes the types of items kept in the catalog and the way in which you can query it to access information about the database structure. The exact structure of the catalog has been simplified, but it is representative of how a catalog works in most relational systems.

The DBMS automatically maintains the system catalog, which contains several tables. The catalog tables you'll consider in this basic introduction are **SYSTABLES** (information about the tables known to SQL), **SYSCOLUMNS** (information about the columns within these tables), and **SYSVIEWS** (information about the views that have been created).

Individual SQL implementations might use different names for these tables. In Oracle, the equivalent tables are named **DBA_TABLES**, **DBA_TAB_COLUMNS**, and **DBA_VIEWS**.

The system catalog is a relational database of its own. Consequently, you can use the same types of queries to retrieve information that you can use to retrieve data in a relational database. You can obtain information about the tables in a relational database, the columns they contain, and the views built on them from the system catalog. The following examples illustrate this process.

Note: Most users need privileges to view system catalog data, so you might not be able to execute these commands. If you are executing the commands shown in the figures, substitute your user name for PRATT to list objects that you own. Your results will differ from those shown in the figures.

ACCESS : In Access, you use the Documenter to obtain the information discussed in
USER : this section, rather than querying the system catalog.

MYSQL : In MySQL, use the SHOW TABLES command to list all the tables in the data-
USER : base. Use the SHOW COLUMNS FROM table command to list all the columns
: in a given table. To list all the columns in the CUSTOMER table, for example,
: the command would be SHOW COLUMNS FROM CUSTOMER.

EXAMPLE 17 : List the name of each table for which the owner (creator of the table) is PRATT.

The command to list the table names owned by PRATT is shown in Figure 6.18. The WHERE clause restricts the tables to only those owned by PRATT.

FIGURE 6.18 Tables owned by PRATT

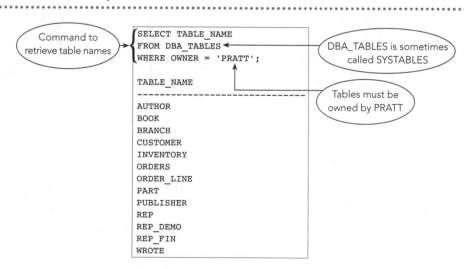

EXAMPLE 18 : List the name of each view owned by PRATT.

This command is similar to the command in Example 17. Rather than TABLE_NAME, the column to be selected is named VIEW_NAME. The command appears in Figure 6.19.

FIGURE 6.19
. . . .
Views owned by
PRATT

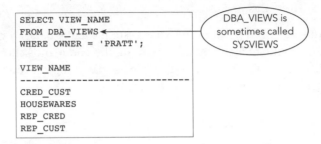

```
SELECT VIEW_NAME
FROM DBA_VIEWS
WHERE OWNER = 'PRATT';


VIEW_NAME
-------------------------------
CRED_CUST
HOUSEWARES
REP_CRED
REP_CUST
```

DBA_VIEWS is sometimes called SYSVIEWS

EXAMPLE 19 : For the CUSTOMER table owned by PRATT, list each column and its data type.

The command for this example appears in Figure 6.20. The columns to select are COLUMN_NAME and DATA_TYPE.

FIGURE 6.20
. . . .
Columns in the
CUSTOMER
table

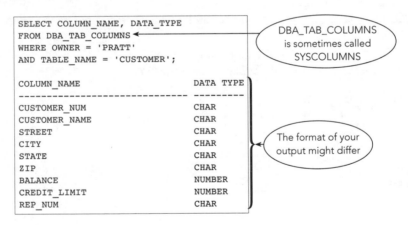

```
SELECT COLUMN_NAME, DATA_TYPE
FROM DBA_TAB_COLUMNS
WHERE OWNER = 'PRATT'
AND TABLE_NAME = 'CUSTOMER';

COLUMN_NAME                              DATA TYPE
--------------------------------------- ---------
CUSTOMER_NUM                             CHAR
CUSTOMER_NAME                            CHAR
STREET                                  CHAR
CITY                                    CHAR
STATE                                   CHAR
ZIP                                     CHAR
BALANCE                                 NUMBER
CREDIT_LIMIT                            NUMBER
REP_NUM                                 CHAR
```

DBA_TAB_COLUMNS is sometimes called SYSCOLUMNS

The format of your output might differ

EXAMPLE 20 : List each table owned by PRATT that contains a column named CUSTOMER_NUM.

This command also queries the SYSCOLUMNS (DBA_TAB_COLUMNS) table. As shown in Figure 6.21, the COLUMN_NAME column is used in the WHERE clause to restrict the rows to those in which the column name is CUSTOMER_NUM.

FIGURE 6.21

· · · · · · · · · · · · · · · · ·

Tables owned by
PRATT with
CUSTOMER_
NUM columns

```
SELECT TABLE_NAME
FROM DBA_TAB_COLUMNS
WHERE OWNER = 'PRATT'
AND COLUMN_NAME = 'CUSTOMER_NUM';


TABLE_NAME
-------------------------------
CUSTOMER
ORDERS
```

When users create, alter, or drop tables or when they create or drop indexes, the DBMS updates the system catalog automatically to reflect these changes. Users should not execute SQL queries to update the catalog directly because this might produce inconsistent information. For example, if a user deletes the CUSTOMER_NUM column in the SYSCOLUMNS table, the system would no longer have any knowledge of this column, which is the CUSTOMER table's primary key, yet all the rows in the CUSTOMER table would still contain a customer number. The system might now treat those customer numbers as names, because as far as the system is concerned, the column named CUSTOMER_NAME is the first column in the CUSTOMER table.

■ Integrity Rules in SQL

An **integrity constraint** is a rule for the data in the database. Examples of integrity constraints in the Premiere Products database are as follows:

- A sales rep's number must be unique.

- The sales rep number for a customer must match the number of a sales rep currently in the database. For example, because there is no sales rep number 11, a customer cannot be assigned to sales rep 11.

- Item classes for parts must be AP, HW, or SG.

If a user enters data in the database that violates any of these integrity constraints, the database will develop serious problems. For example, two sales reps with the same number, a customer with a nonexistent sales rep, or a part in a nonexistent item class would compromise the integrity of data in the database. To manage these types of problems, the DBMS provides **integrity support**, the process of specifying integrity constraints for a database that the DBMS will enforce. SQL has clauses to support three types of integrity constraints that you can specify within a CREATE TABLE or an ALTER TABLE command. The only difference between these two commands is that an ALTER TABLE command is followed by the word ADD to indicate that you are adding the constraint to the list of existing constraints. To change an integrity constraint after it has been created, just enter the new constraint, which immediately takes the place of the original.

The types of constraints supported in SQL are primary keys, foreign keys, and legal values. In most cases, you specify a table's primary key when you create the table. To add a primary key after creating a table, you can use the **ADD PRIMARY KEY** clause of the ALTER TABLE command. For example, to indicate that REP_NUM is the primary key for the REP table, the ALTER TABLE command would be:

```
ALTER TABLE REP
ADD PRIMARY KEY (REP_NUM);
```

The PRIMARY KEY clause is PRIMARY KEY followed by the column name that makes up the primary key in parentheses. If the primary key contains more than one column, use commas to separate the column names.

ACCESS USER : To specify a table's primary key in Access, open the table in Design view, select the column(s) that make up the primary key, and then click the Primary Key button on the toolbar.

A **foreign key** is a column in one table whose values match the primary key in another table. (One example is the CUSTOMER_NUM column in the ORDERS table. Values in this column are required to match those of the primary key in the CUSTOMER table.)

EXAMPLE 21 : Specify the CUSTOMER_NUM column in the ORDERS table as a foreign key that must match the CUSTOMER table.

When a table contains a foreign key, you identify it using the **ADD FOREIGN KEY** clause of the ALTER TABLE command. In this clause, you specify *both* the column that is a foreign key *and* the table it matches. The general form for assigning a foreign key is FOREIGN KEY, the column name(s) of the foreign key, the **REFERENCES** clause, and then the table name that the foreign key must match as shown in Figure 6.22.

FIGURE 6.22   Adding a foreign key to an existing table

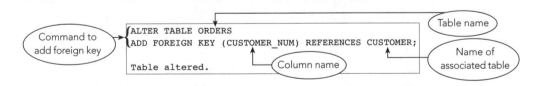

ACCESS USER : To specify a foreign key in Access, open the Relationships window, relate the corresponding tables, and then indicate that you want Access to enforce referential integrity.

After creating a foreign key, the DBMS will reject any update that violates the foreign key constraint. For example, the DBMS rejects the INSERT command shown in

Figure 6.23 because it attempts to add an order for which the customer number (850) does not match any customer in the CUSTOMER table. The DBMS also rejects the DELETE command in Figure 6.23 because it attempts to delete customer number 148; rows in the ORDERS table for which the customer number is 148 would no longer match any row in the CUSTOMER table.

FIGURE 6.23 Violating foreign key constraints

```
INSERT INTO ORDERS
VALUES
('21625','23-OCT-2007','850');

INSERT INTO ORDERS
*
ERROR at line 1:
ORA-02291: integrity constraint (SYSTEM.SYS_C002865) violated - parent key not found


DELETE FROM CUSTOMER
WHERE CUSTOMER_NUM = '148';
DELETE FROM CUSTOMER
*
ERROR at line 1:
ORA-02292: integrity constraint (SYSTEM.SYS_C002865) violated - child record found
```

Attempting to add a row with a nonexistent customer number

Attempting to delete a customer with associated orders

Note that the error messages shown in Figure 6.23 include the words "parent" and "child." When you specify a foreign key, the table containing the foreign key is the **child** and the table referenced by the foreign key is the **parent**. For example, the CUSTOMER_NUM column in the ORDERS table is a foreign key that references the CUSTOMER table. For this foreign key, the CUSTOMER table is the parent and the ORDERS table is the child. The first error message indicates that there is no parent for the order (there is no customer number 850). The second error message indicates that there are child records (rows) for customer 148 (customer 148 has orders). The DBMS rejects both updates because they violate referential integrity.

EXAMPLE 22 : Specify the valid item classes for the PART table as AP, HW, and SG.

You use the **CHECK** clause of the ALTER TABLE command to ensure that only legal values satisfying a particular condition are allowed in a given column. The general form of the CHECK clause is simply the word CHECK followed by a condition. If a user enters data that violates the condition, the DBMS rejects the update automatically. For example, to ensure that the only legal values for item class are AP, HW, or SG, use one of the following versions of the CHECK clause:

```
CHECK (CLASS IN ('AP', 'HW', 'SG') )
```

or

```
CHECK (CLASS = 'AP' OR CLASS = 'HW' OR CLASS = 'SG')
```

The ALTER TABLE command shown in Figure 6.24 uses the first version of the CHECK clause.

```
ALTER TABLE PART
ADD CHECK (CLASS IN ('AP', 'HW', 'SG') );

Table altered.
```
Valid classes are AP, HW, and SG

ACCESS : In Access, you specify a validation rule in Design view rather than using
USER : the CHECK clause.

Now the DBMS will reject the update shown in Figure 6.25 because the command attempts to change the item class to XX, which is an illegal value.

```
UPDATE PART
SET CLASS = 'XX'
WHERE PART_NUM = 'AT94';
UPDATE PART
*
ERROR at line 1:
ORA-02290: check constraint (SYSTEM.SYS_C002868) violated
```
XX class is invalid

In this chapter, you learned about the purpose, creation, use, and benefits of views. Then you examined the features of SQL that relate to security, including granting and revoking various privileges that control the types of activities users of the database need to perform. You learned about the purpose, advantages, and disadvantages of using indexes, and how to create and drop indexes. You learned how to obtain information from the system catalog. Finally, you learned about the importance of integrity constraints and how to create them. In the next chapter, you'll learn how to use SQL commands to create and format reports.

◼ SUMMARY

- A view contains data that is derived from existing base tables when users attempt to access the view.

- To define a view, use the CREATE VIEW command, which includes a defining query that describes the portion of the database included in the view. When a user retrieves data from the view, the DBMS merges the query entered by the user with the defining query and produces the query that SQL actually executes.

- Views provide data independence, allow database access control, and simplify the database structure for users.

- You cannot update views that involve statistics and views with joins of non-primary key columns. Updates for these types of views must be made in the base table.

- Use the DROP VIEW command to delete a view.

- Use the GRANT command to give users access privileges to data in the database.

- Use the REVOKE command to terminate previously granted privileges.

- You can create and use an index to make data retrieval more efficient.

- Use the CREATE INDEX command to create an index. Use the CREATE UNIQUE INDEX command to enforce a rule so that only unique values are allowed in a non-primary key column.

- Use the DROP INDEX command to delete an index.

- The DBMS, not the user, chooses which index to use to accomplish a given task.

- The DBMS maintains information about the tables, columns, indexes, and other system elements in the system catalog (catalog) or data dictionary. Information about tables is kept in the SYSTABLES table, information about columns is kept in the SYSCOLUMNS table, and information about views is kept in the SYSVIEWS table. In Oracle, these same tables are named DBA_TABLES, DBA_TAB_COLUMNS, and DBA_VIEWS.

- Use the SELECT command to obtain information from the system catalog. The DBMS updates the system catalog automatically whenever changes are made to the database.

- Integrity constraints are rules that the data in the database must follow to ensure that only legal values are accepted in specified columns and that primary and foreign key values match between tables. To specify a general integrity constraint, use the CHECK clause. You usually specify primary key constraints when you create a table, but you can specify them later using the ADD PRIMARY KEY clause. To specify a foreign key, use the ADD FOREIGN KEY clause.

■ KEY TERMS

ADD FOREIGN KEY
ADD PRIMARY KEY
base table
catalog
CHECK
child
CREATE INDEX
CREATE UNIQUE INDEX
CREATE VIEW

data dictionary
database administration
database administrator
DBA_TAB_COLUMNS
DBA_TABLES
DBA_VIEWS
defining query
DROP INDEX
DROP VIEW

foreign key
GRANT
index
integrity constraint
integrity support
parent
REFERENCES
REVOKE
row-and-column subset view

security
SYSCOLUMNS
SYSTABLES
system catalog
SYSVIEWS
unique index
view
WITH GRANT OPTION

■ REVIEW QUESTIONS

1. What is a view?

2. Which command defines a view?

3. What is a defining query?

4. What happens when a user retrieves data from a view?

5. What are three advantages of using views?

6. Which types of views cannot be updated?

7. Which command deletes a view?

8. Which command gives users access privileges to various portions of the database?

9. Which command terminates previously granted privileges?

10. What is the purpose of an index?

11. How do you create an index? How do you create a unique index? What is the difference between an index and a unique index?

12. Which command deletes an index?

13. Does the DBMS or the user make the choice of which index to use to accomplish a given task?

14. Describe the information the DBMS maintains in the system catalog. What are the generic names for three tables in the catalog and their corresponding names in Oracle?

15. Which command do you use to obtain information from the system catalog?

16. How is the system catalog updated?

17. What are integrity constraints?

18. How do you specify a general integrity constraint?

19. When are primary key constraints usually specified? Can you specify them after creating a table? How?

20. How do you specify a foreign key in Oracle?

◼ EXERCISES (Premiere Products)

Use SQL to make the following changes to the Premiere Products database (see Figure 1.2 in Chapter 1). For any exercises that use commands not supported by your version of SQL, write the command to accomplish the task. Use the Notes at the end of Chapter 2 to print your output if directed to do so by your instructor.

1. Define a view named MAJOR_CUSTOMER. It consists of the customer number, name, balance, credit limit, and rep number for every customer whose credit limit is $10,000 or less.
   a. Write and execute the CREATE VIEW command to create the MAJOR_CUSTOMER view.
   b. Write and execute the command to retrieve the customer number and name of each customer in the MAJOR_CUSTOMER view with a balance that exceeds the credit limit.
   c. Write and execute the query that the DBMS actually executes.
   d. Does updating the database through this view create any problems? If so, what are they? If not, why not?

2. Define a view named PART_ORDER. It consists of the part number, description, price, order number, order date, number ordered, and quoted price for all order lines currently on file.
   a. Write and execute the CREATE VIEW command to create the PART_ORDER view.
   b. Write and execute the command to retrieve the part number, description, order number, and quoted price for all orders in the PART_ORDER view for parts with quoted prices that exceed $100.
   c. Write and execute the query that the DBMS actually executes.

   d. Does updating the database through this view create any problems? If so, what are they? If not, why not?

3. Define a view named ORDER_TOTAL. It consists of the order number and order total for each order currently on file. (The order total is the sum of the number of units ordered times the quoted price on each order line for each order.) Sort the rows by order number. Use TOTAL_AMOUNT as the name for the order total.
   a. Write and execute the CREATE VIEW command to create the ORDER_TOTAL view.
   b. Write and execute the command to retrieve the order number and order total for only those orders totaling more than $1,000.
   c. Write and execute the query that the DBMS actually executes.
   d. Does updating the database through this view create any problems? If so, what are they? If not, why not?

4. Write, but do not execute, the SQL commands to grant the following privileges:
   a. User Ashton must be able to retrieve data from the PART table.
   b. Users Kelly and Morgan must be able to add new orders and order lines.
   c. User James must be able to change the price for all parts.

d. User Danielson must be able to delete customers.
e. All users must be able to retrieve each customer's number, name, street, city, state, and zip code.
f. User Perez must be able to create an index on the ORDERS table.
g. User Washingon must be able to change the structure of the PART table.
h. User Grinstead must have all privileges on the ORDERS, ORDER_LINE, and PART tables.
i. User Webb must be permitted to access any data concerning housewares but is restricted from accessing data concerning any other parts.

5. Write, but do not execute, the SQL command to revoke all privileges from user Ashton.

6. Perform the following tasks:
   a. Create an index named PART_INDEX1 on the PART_NUM column in the ORDER_LINE table.
   b. Create an index named PART_INDEX2 on the CLASS column in the PART table.
   c. Create an index named PART_INDEX3 on the CLASS and WAREHOUSE columns in the PART table.
   d. Create an index named PART_INDEX4 on the CLASS and WAREHOUSE columns in the PART table. List item classes in descending order.

7. Delete the index named PART_INDEX3.

8. Write the SQL commands to obtain the following information from the system catalog for the objects that you own. Do not execute these commands unless your instructor asks you to do so.
   a. List every table.
   b. List every column in the PART table and its associated data type.
   c. List every table that contains a column named ORDER_NUM.
   d. List the name of every view in the system that you created.
   e. List the table name, column name, and data type for the columns named STREET, CITY, STATE, and ZIP. Order the results by table name within column name. (That is, column name is the major sort key and table name is the minor sort key.)

9. Add ORDER_NUM as a foreign key in the ORDER_LINE table.

10. Ensure that the only values entered into the CREDIT_LIMIT column are 5000, 7500, 10000, and 15000.

■EXERCISES (Henry Books)

Use SQL to make the following changes to the Henry Books database (see Figures 1.4 through 1.7 in Chapter 1). For any exercises that use commands not supported by your version of SQL, write the command to accomplish the task. Use the Notes at the end of Chapter 2 to print your output if directed to do so by your instructor.

1. Define a view named SCRIBNER. It consists of the book code, title, type, and price for every book published by the publisher whose code is SC.
   a. Write and execute the CREATE VIEW command to create the SCRIBNER view.
   b. Write and execute the command to retrieve the book code, title, and price for every book with a price of less than $15.
   c. Write and execute the query that the DBMS actually executes.
   d. Does updating the database through this view create any problems? If so, what are they? If not, why not?

2. Define a view named NON_PAPERBACK. It consists of the book code, title, publisher name, and price for every book that is not available in paperback.
   a. Write and execute the CREATE VIEW command to create the NON_PAPERBACK view.
   b. Write and execute the command to retrieve the book title, publisher name, and price for every book in the NON_PAPERBACK view with a price of less than $20.
   c. Write and execute the query that the DBMS actually executes.

   d. Does updating the database through this view create any problems? If so, what are they? If not, why not?

3. Define a view named BOOK_INVENTORY. It consists of the branch number and the total number of books on hand for each branch. Use UNITS as the name for the count of books on hand. Group and order the rows by branch number.
   a. Write and execute the CREATE VIEW command to create the BOOK_INVENTORY view.
   b. Write and execute the command to retrieve the branch number and units for each branch having more than 25 books on hand.
   c. Write and execute the query that the DBMS actually executes.
   d. Does updating the database through this view create any problems? If so, what are they? If not, why not?

4. Write, but do not execute, SQL commands to grant the following privileges:
   a. User Rodriguez must be able to retrieve data from the BOOK table.
   b. Users Gomez and Liston must be able to add new books and publishers to the database.

c. Users Andrews and Zimmer must be able to change the price of any book.

d. All users must be able to retrieve the book title, book code, and book price for every book.

e. User Golden must be able to add and delete publishers.

f. User Andrews must be able to create an index for the BOOK table.

g. Users Andrews and Golden must be able to change the structure of the AUTHOR table.

h. User Golden must have all privileges on the BRANCH, BOOK, and INVENTORY tables.

i. User Duvall must be able to change the units on hand for books in branch number 2 but must be unable to access data in any other branch.

5. Write, but do not execute, the SQL command to revoke all privileges from user Andrews.

6. Create the following indexes:

a. Create an index named BOOK_INDEX1 on the TITLE column in the BOOK table.

b. Create an index named BOOK_INDEX2 on the TYPE column in the BOOK table.

c. Create an index named BOOK_INDEX3 on the CITY and PUBLISHER_NAME columns in the PUBLISHER table.

7. Delete the index named BOOK_INDEX3; it is no longer necessary.

8. Write the SQL commands to obtain the following information from the system catalog for the objects that you own. Do not execute these commands unless your instructor asks you to do so.

a. List every table.

b. List every column in the PUBLISHER table and its associated data type.

c. List every table that contains a column named PUBLISHER_CODE.

d. List the name of every view.

e. List the table name, column name, and data type for the columns named BOOK_CODE, TITLE, and PRICE. Order the results by table name within column name. (That is, column name is the major sort key and table name is the minor sort key.)

9. Add PUBLISHER_CODE as a foreign key in the BOOK table.

10. Ensure that the PAPERBACK column in the BOOK table can accept only values of Y or N.

■EXERCISES (Alexamara Marina Group)

Use SQL to make the following changes to the Alexamara Marina Group database (Figures 1.8 through 1.12 in Chapter 1). For any exercises that use commands not supported by your version of SQL, write the command to accomplish the task. Use the Notes at the end of Chapter 2 to print your output if directed to do so by your instructor.

1. Define a view named LARGE_SLIP. It consists of the marina number, slip number, rental fee, boat name, and owner number for every slip whose length is 40 feet.
   a. Write and execute the CREATE VIEW command to create the LARGE_SLIP view.
   b. Write and execute the command to retrieve the marina number, slip number, rental fee, and boat name for every slip with a rental fee of $3,800.00 or more.
   c. Write and execute the query that the DBMS actually executes.
   d. Does updating the database through this view create any problems? If so, what are they? If not, why not?

2. Define a view named RAY_4025. It consists of the marina number, slip number, length, rental fee, boat name, and owner's last name for every slip in which the boat type is Ray 4025.
   a. Write and execute the CREATE VIEW command to create the RAY_4025 view.
   b. Write and execute the command to retrieve the marina number, slip number, rental fee, boat name, and owner's last name for every slip in the RAY_4025 view with a rental fee of less than $4,000.00.
   c. Write and execute the query that the DBMS actually executes.

   d. Does updating the database through this view create any problems? If so, what are they? If not, why not?

3. Define a view named SLIP_FEES. It consists of two columns: The first is the slip length, and the second is the average fee for all slips in the MARINA_SLIP table that have that length. Use AVERAGE_FEE as the name for the average fee. Group and order the rows by slip length.
   a. Write and execute the CREATE VIEW command to create the SLIP_FEES view.
   b. Write and execute the command to retrieve the slip length and average fee for each length for which the average fee is less than $3,500.00.
   c. Write and execute the query that the DBMS actually executes.
   d. Does updating the database through this view create any problems? If so, what are they? If not, why not?

4. Write, but do not execute, the SQL commands to grant the following privileges:
   a. User Oliver must be able to retrieve data from the MARINA_SLIP table.
   b. Users Crandall and Perez must be able to add new owners and slips to the database.
   c. Users Johnson and Klein must be able to change the rental fee of any slip.

d. All users must be able to retrieve the length, boat name, and owner number for every slip.

e. User Klein must be able to add and delete service categories.

f. User Adams must be able to create an index on the SERVICE_REQUEST table.

g. Users Adams and Klein must be able to change the structure of the MARINA_SLIP table.

h. User Klein must have all privileges on the MARINA, OWNER, and MARINA_SLIP tables.

5. Write, but do not execute, the SQL command to revoke all privileges from user Klein.

6. Create the following indexes:

a. Create an index named BOAT_INDEX1 on the OWNER_NUM column in the MARINA_SLIP table.

b. Create an index named BOAT_INDEX2 on the BOAT_NAME column in the MARINA_SLIP table.

c. Create an index named BOAT_INDEX3 on the LENGTH and BOAT_NAME columns in the MARINA_SLIP table. List the lengths in descending order.

7. Delete the index named BOAT_INDEX3; it is no longer necessary.

8. Write the SQL commands to obtain the following information from the system catalog for the objects that you own. Do not execute these commands unless your instructor specifically asks you to do so.

a. List every table.

b. List every column in the MARINA_SLIP table and its associated data type.

c. List every table that contains a column named OWNER_NUM.

d. List the name of every view.

9. Add OWNER_NUM as a foreign key in the MARINA_SLIP table.

10. Ensure that the LENGTH column in the MARINA_SLIP table can accept only values of 25, 30, or 40.

CHAPTER **7**

Reports

OBJECTIVES

- Understand how to use functions in queries

- Use the UPPER and LOWER functions with character data

- Use the ROUND and FLOOR functions with numeric data

- Add a specific number of months or days to a date

- Calculate the number of days between two dates

- Use concatenation in a query

- Create a view for a report

- Create a query for a report

- Change column headings and formats in a report

- Add a title to a report

- Group data in a report

- Include totals and subtotals in a report

- Send a report to a file that can be printed

Introduction

You have already seen functions that apply to groups (SUM, AVG, and so on). In this chapter, you will learn to use functions that apply to values in individual rows. Specifically, you will see how to use the UPPER and LOWER functions with character data. You will also learn how to use the ROUND and FLOOR functions with numeric data and perform calculations with dates.

In addition to the various functions that you have used in this text, some SQL implementations include commands that you can use to format a report. In this chapter, you will see how to use Oracle commands to format reports.

In a sense, you can think of the output of a SELECT command as a very simple report that follows a rigid format. This format is sufficient if you are concerned only about the particular data that appears in the result but not about its format. If you need to generate a formatted report, however, you can't use a SELECT command. Fortunately, you can use other SQL commands to format query output in a variety of ways, and then you can run the query again later with the results displayed in the desired format.

■ Using Functions

You have already used aggregate functions to perform calculations based on groups of records. For example, SUM(BALANCE) calculates the sum of the balances on all records that satisfy the condition in the WHERE clause. If you use a GROUP BY clause, the sum will be calculated for each record in a group.

SQL also includes functions that affect single records. Some functions affect characters and others let you manipulate numeric data. The list of SQL functions varies from one implementation of SQL to another. This section will illustrate some common functions. For additional information about the functions available to you, consult the documentation for your implementation of SQL.

Character Functions

SQL includes several functions that affect character data. Example 1 illustrates the use of one of the them, the UPPER function.

EXAMPLE 1 : List the rep number and last name for each sales rep. The last name should appear in uppercase letters.

The **UPPER** function changes a value to uppercase letters; for example, the function UPPER(LAST_NAME) would change the last name Kaiser to KAISER. The item in parentheses (LAST_NAME) is called the **argument** for the function. The value produced by the function is the result of converting all lowercase letters in the last name stored in the LAST_NAME column to uppercase letters. The query and its results are shown in Figure 7.1.

FIGURE 7.1   Using the UPPER function to convert character data to uppercase letters

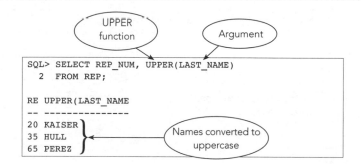

You can use functions in WHERE clauses as well. For example, the condition UPPER(LAST_NAME) = 'KAISER' would be true for names like Kaiser, KAISER, and KaIsER, because the result of applying the UPPER function to any of these values would be KAISER.

To convert a value to lowercase letters, you can use the **LOWER** function.

ACCESS   : In Access, the **UCASE** function converts a value to uppercase letters and
    USER   : the LCASE function converts a value to lowercase letters. For example, if
            : the value stored in the LAST_NAME column is Kaiser, UCASE(LAST_NAME)
            : would be KAISER and LCASE(LAST_NAME) would be kaiser.

Number Functions

SQL also includes functions that affect numeric data. The **ROUND** function, which rounds values to a specified number of decimal places, is illustrated in Example 2.

EXAMPLE 2   : List the part number and price for all parts. Round the price to the nearest
            : whole dollar amount.

A function can have more than one argument. The ROUND function, which rounds a numeric value to a desired number of decimal places, has two arguments. The first argument is the value to be rounded; the second argument indicates the number of decimal places to which to round the result. For example, ROUND(PRICE,0) will round the values in the PRICE column to zero decimal places (a whole number). If a price is 24.95, the result will be 25. If the price is 24.25, on the other hand, the result will be 24. Figure 7.2 shows the query and results to round values in the PRICE column to zero decimal places. The computed column ROUND(PRICE,0) is named ROUNDED_PRICE.

FIGURE 7.2   Using the ROUND function to round numeric values

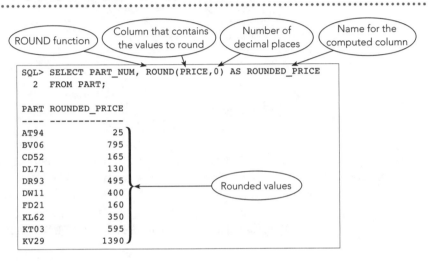

Rather than rounding (using the ROUND function), you might need to simply truncate (remove) everything to the right of the decimal point. To do so, use the **FLOOR** function, which has only one argument. If a price is 24.95, for example, ROUND(PRICE,0) would result in 25, whereas FLOOR(PRICE) would result in 24.

Working with Dates

SQL uses functions and calculations for manipulating dates. To add a specific number of months to a date, you can use the **ADD_MONTHS** function as illustrated in Example 3.

EXAMPLE 3   For each order, list the order number and the date that is two months after the order date. Call this date FUTURE_DATE.

The ADD_MONTHS function has two arguments. The first argument is the date to which you wish to add a specific number of months, and the second argument is the number of months. To add two months to the order date, for example, the expression is ADD_MONTHS(ORDER_DATE,2) as illustrated in Figure 7.3.

FIGURE 7.3   Using the ADD_MONTHS function to add months to a date

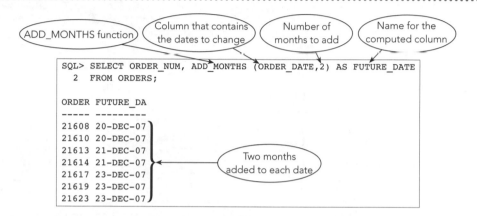

ACCESS USER : To add a number of months to a date in Access, use the **DATEADD** function. This function has three arguments. The first argument includes the interval of time to be added; the letter "m" indicates that months will be added. The second argument includes the number of intervals to be added. The third argument includes the date to be manipulated. For example, to add two months to the dates stored in the ORDER_DATE column, the appropriate function would be DATEADD("m", 2, ORDER_DATE).

MYSQL USER : To add a number of months to a date in MySQL, use the **ADDDATE** function. This function has two arguments. The first argument includes the date to be manipulated; the second argument includes the interval to be added. The second argument consists of the word INTERVAL followed by the appropriate interval. For example, to add two months to the dates stored in the ORDER_DATE column, the appropriate function would be ADDDATE(ORDER_DATE, INTERVAL 2 MONTH).

EXAMPLE 4 : For each order, list the order number and the date that is seven days after the order date. Call this date FUTURE_DATE.

To add a specific number of days to a date, you do not need a function. You can add the number of days to the order date as illustrated in Figure 7.4.

FIGURE 7.4   Adding days to dates

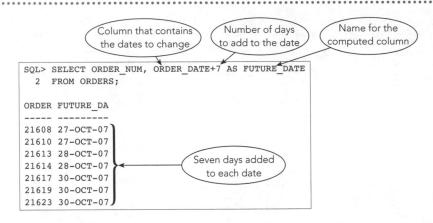

Column that contains the dates to change

Number of days to add to the date

Name for the computed column

```
SQL> SELECT ORDER_NUM, ORDER_DATE+7 AS FUTURE_DATE
  2  FROM ORDERS;


ORDER FUTURE_DA
----- ---------
21608 27-OCT-07
21610 27-OCT-07
21613 28-OCT-07
21614 28-OCT-07
21617 30-OCT-07
21619 30-OCT-07
21623 30-OCT-07
```

Seven days added to each date

ACCESS USER : To add a number of days to a date in Access, use the DATEADD function with the letter "d" (day) as the time interval. For example, to add seven days to the dates stored in the ORDER_DATE column, the appropriate function would be DATEADD("d", 7, ORDER_DATE).

MYSQL USER : To add a number of days to a date in MySQL, you need to use the ADDDATE function with the DAY interval rather than simply performing the arithmetic shown in Figure 7.4. To add seven days to the dates stored in the ORDER_DATE column, the appropriate function would be ADDDATE(ORDER_DATE, INTERVAL 7 DAY).

EXAMPLE 5 : For each order, list the order number, today's date, the order date, and the number of days between the order date and today's date. Call today's date TODAYS_DATE and call the number of days between the order date and today's date DAYS_PAST.

You can use the special word **SYSDATE** to obtain today's date, as shown in Figure 7.5. The command in the figure uses SYSDATE to display today's date and also uses SYSDATE in a computation to determine the number of days between the order date and today's date. The values for DAYS_PAST include decimal places. You could remove these decimal places by using the ROUND or FLOOR functions, if desired.

FIGURE 7.5   Calculating the number of days between two dates

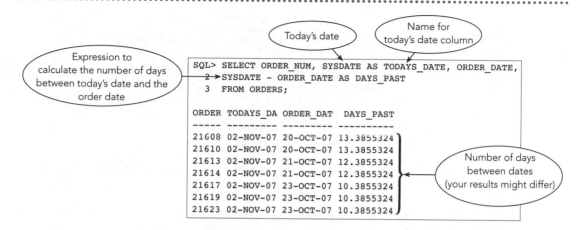

ACCESS USER

In Access, use the **DATE** function to obtain today's date, rather than SYSDATE. This function has no arguments, so you would simply write DATE() in place of SYSDATE.

MYSQL USER

In MySQL, use the **CURDATE** function to obtain today's date, rather than SYSDATE. This function has no arguments, so you would simply write CURDATE() in place of SYSDATE.

■ Concatenating Columns

Sometimes you need to **concatenate**, or combine, two or more character columns into a single expression when displaying them in a report; the process is called **concatenation**. To concatenate columns, you type two vertical lines (| |) between the column names, as illustrated in Example 6.

EXAMPLE 6   List the number and name of each sales rep. The name should be a concatenation of the FIRST_NAME and LAST_NAME columns.

The command and its results appear in Figure 7.6. The vertical lines between the column names in the SELECT clause indicate a concatenation. Notice that the first and last names of each sales rep now appear in a single column in the query results.

FIGURE 7.6   Concatenating two columns

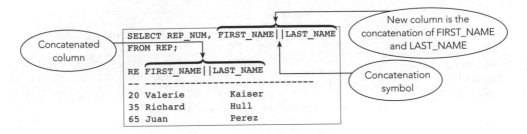

ACCESS **: In Access, use the & symbol to concatenate columns. The corresponding
USER : query in Access would be:**

```
SELECT REP_NUM, FIRST_NAME&LAST_NAME
FROM REP;
```

MYSQL **: In MySQL, use the CONCAT function to concatenate columns. The corre-
USER : sponding query would be:**

```
SELECT REP_NUM, CONCAT(FIRST_NAME, LAST_NAME)
FROM REP;
```

When the first name doesn't include sufficient characters to fill the width of the column (as determined by the number of characters specified in the CREATE TABLE command), SQL inserts extra spaces. To remove these extra spaces, you use the **RTRIM** (right trim) function. When you apply this function to the value in a column, SQL displays the original value and removes any spaces inserted at the end of the value. Figure 7.7 shows the query and output with the extra spaces removed. For sales rep 20, for example, this command trims the first name to "Valerie," concatenates it with a single space, and then concatenates the last name "Kaiser."

FIGURE 7.7

Concatenating
two columns
using the RTRIM
function

```
SELECT REP_NUM, RTRIM(FIRST_NAME)||' '||RTRIM(LAST_NAME)
FROM REP;

RE RTRIM(FIRST_NAME)||' '||RTRIM(LA
-- --------------------------------
20 Valerie Kaiser
35 Richard Hull          Spaces removed
65 Juan Perez
```

QUESTION Why is it necessary to insert a single space character in single quotation
marks in the query?

ANSWER Without the space character, there would be no space between the first
and last names. The name of sales rep 20, for example, would be displayed as "ValerieKaiser."

ACCESS In Access, it is not necessary to trim the columns because Access will trim
USER them automatically. To insert the space between values, the query in
Access would be:

```
SELECT REP_NUM, FIRST_NAME&' '&LAST_NAME
FROM REP;
```

MYSQL In MySQL, it is not necessary to trim the columns because MySQL will trim
USER them automatically. To insert the space between values, the query in
MySQL would be:

```
SELECT REP_NUM, CONCAT(FIRST_NAME,' ',LAST_NAME)
FROM REP;
```

ACCESS Access does not support the report-formatting commands discussed in
USER the rest of this chapter. Instead of using these commands, you would create and save a query containing the data you want to include in the report.
Then you would use the Report Wizard or Report Design view to create a
report based on the saved query.

MYSQL MySQL does not support the report-formatting commands discussed in
USER the rest of this chapter.

■ Creating and Using Scripts

When creating views and also when entering report formatting commands, it's a good idea
to save the commands in script files for future use. Otherwise, you must reenter the commands every time you want to produce the same report.

Note: When you create a report, you will typically create a view and three files: the script to create the view, the script to format the report, and the report output. The scripts and the report all have the extension .sql. To manage these files, you can save the script that creates the view with the view's name, the name of the view plus _FORMAT for the script that creates the report, and the name of the view plus _OUTPUT for the report output. For example, for a view named SLSR_REPORT, you would name the script that creates the view SLSR_REPORT.sql, the script that formats the report SLSR_REPORT_FORMAT.sql, and the file that contains the report output SLSR_REPORT_OUTPUT.sql.

Recall that in SQL*Plus, you can use an editor such as Notepad to create script files by typing the EDIT command and the name of the file you want to create. Oracle assigns the file the extension .sql automatically. Then you type the command(s), save the file, and close the editor. To run the command(s) in the file from SQL*Plus, type @ (the "at" symbol) followed by the name of the file. For example, to run a script file named SLSR_REPORT, you would type @SLSR_REPORT. (If you saved the script file in a folder other than the default folder for your DBMS, you would need to include the full path to the file as part of the filename.) After you press the Enter key, SQL executes the command(s) saved in the file.

In SQL*Plus Worksheet, you can create a script file by typing the command(s) in the upper pane of the SQL*Plus Worksheet window, selecting the Save Input As command on the File menu, and specifying the name and location for the file. Oracle assigns the file the extension .sql automatically. To run the command(s) in the file, select the Open command on the File menu, select the file, click the Open button, and then click the Execute button.

Note: You can save the script file to any storage location by including a drive and/or folder designation. For example, to create the file named SLSR_REPORT in a folder named ORACLE on drive A, the file would be A:\ORACLE\SLSR_REPORT.

Note: In some SQL implementations, you must type a slash at the end of the command(s) and press the Enter key for the script to run correctly.

Note: If you receive an error message that your input is truncated when you try to run a script, edit your script to add a hard return after the slash.

Saving your commands in a script has another advantage—doing so allows you to develop your report in stages. You can create a file with an initial set of commands to format the report, and then see what the report looks like. If necessary, you can change the commands in the script to improve the report's appearance and then view the report again. You can make additional changes and view the report until you are satisfied with its appearance.

■ Running the Query for the Report

The data for a report can come from either a table or a view. Using a view is preferable to using a table, particularly if the report involves data from more than one table.

The report you will create in this chapter displays data from the REP and CUSTOMER tables. The report also concatenates two column names, as illustrated in Example 6. Example 7 creates the view for the report.

EXAMPLE 7   Create a script file named SLSR_REPORT.sql that defines a view named SLSR_REPORT with five columns that you'll use for the report. Name the first column SLSR; it is the concatenation of the sales rep number, first name, and last name for each sales rep. Insert a hyphen between the sales rep number and name, separate the first and last names with a single space, and trim the values. (For example, the resulting value in the SLSR column for the first sales rep should be "20-Valerie Kaiser.") Name the second column CUST; it is the concatenation of the customer number and the customer name. Insert a hyphen between the customer number and name and trim the name. Name the third column BAL; it contains the balance. Name the fourth column CRED; it contains the credit limit. Name the fifth column AVAIL; it contains the available credit (CREDIT_LIMIT - BALANCE) for each customer. Run the script file to create the view.

The script containing the command to create this view appears in Figure 7.8. The first line indicates that the view will consist of five columns named SLSR, CUST, BAL, CRED, and AVAIL. The second line contains the expression for the SLSR column (the rep number, a hyphen, a trimmed version of the first name, a space, and a trimmed version of the last name). The third line contains an expression for the CUST column (the customer number, a hyphen, and a trimmed version of the customer name). The rest of the command lists the tables from which to select data and joins the REP and CUSTOMER tables in a WHERE clause.

FIGURE 7.8   Command to create the view for the report

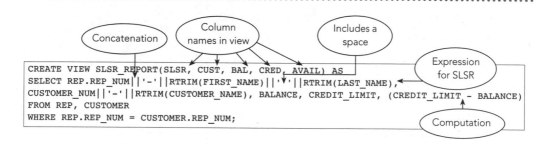

```
CREATE VIEW SLSR_REPORT(SLSR, CUST, BAL, CRED, AVAIL) AS
SELECT REP.REP_NUM||'-'||RTRIM(FIRST_NAME)||' '||RTRIM(LAST_NAME),
CUSTOMER_NUM||'-'||RTRIM(CUSTOMER_NAME), BALANCE, CREDIT_LIMIT, (CREDIT_LIMIT - BALANCE)
FROM REP, CUSTOMER
WHERE REP.REP_NUM = CUSTOMER.REP_NUM;
```

When you run the script using the appropriate method for the version of SQL you are using, you'll see a message that the view has been created.

■ Creating the Data for the Report

To produce a report, you must run a SELECT command to create the data to use in the report. As a basis for this discussion on report formatting, you will use the query described in Example 8.

EXAMPLE 8 : List all the data in the SLSR_REPORT view. Order the rows by the SLSR and
: CUST columns.

The SELECT command and the query results appear in Figure 7.9. The values in the SLSR column consist of the sales rep number, a hyphen, and a trimmed version of the sales rep's name. Similarly, the values in the CUST column consist of the customer number, a hyphen, and a trimmed version of the customer name. Notice that the rows in the output are wider than the screen, causing each row to be displayed on two lines.

In the next examples, you will modify the format of the report to correct these problems, change the column headings, add a title, change the format of the numbers, and add totals and subtotals. When you are finished, each row will appear on a single line.

FIGURE 7.9 Data in the SLSR_REPORT view

```
SELECT *
FROM SLSR_REPORT
ORDER BY SLSR, CUST;

SLSR                                    CUST                                              BAL        CRED
-------------------------------         ----------------------------------------    ----------   --------
        AVAIL
----------
20-Valerie Kaiser                       148-Al's Appliance and Sport                     6550       7500
      950

20-Valerie Kaiser                       524-Kline's                                     12762      15000
     2238

20-Valerie Kaiser                       842-All Season                                   8221       7500
     -721

35-Richard Hull                         282-Brookings Direct                            431.5      10000
   9568.5

35-Richard Hull                         408-The Everything Shop                        5285.25      5000
    -285.25

35-Richard Hull                         687-Lee's Sport and Appliance                    2851       5000
     2149

35-Richard Hull                         725-Deerfield's Four Seasons                      248       7500
     7252

65-Juan Perez                           356-Ferguson's                                   5785       7500
     1715

65-Juan Perez                           462-Bargains Galore                              3412      10000
     6588

65-Juan Perez                           608-Johnson's Department Store                   2106      10000
     7894

10 rows selected.
```

Selects all columns from the view

Orders rows by SLSR and CUST

SLSR is a concatenation of sales rep number, first name, and last name

Rows extend over multiple lines (your output might differ)

■ Changing Column Headings

The column headings shown in Figure 7.9 are not very descriptive of the columns' contents. You can change the headings to improve readability, as illustrated in Example 9.

: Change the column headings in the report so they are more descriptive of
: the columns' contents.

To change a column heading, type the **COLUMN** command followed by the name of
the column heading you want to change. Then use the **HEADING** clause to assign a new
heading. If you want to display the heading on two lines, separate the two portions of the
heading with a single vertical line (|). The commands to change the column headings
appear in Figure 7.10.

FIGURE 7.10   Script with commands to change the column headings

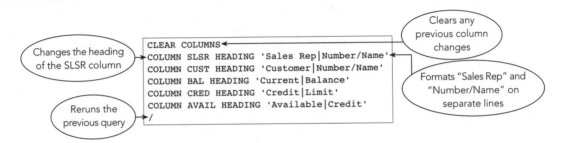

Note: If you follow the suggested naming conventions for scripts, you would name the
script shown in Figure 7.10 SLSR_REPORT_FORMAT.sql. To execute the commands,
you would run the script file.

The **CLEAR COLUMNS** command clears any previous column changes that you
made to column headings or formats in the current work session. The COLUMN commands change the column headings. The slash (/) on the last line reruns the last query and
displays the data with the new column headings. Running the script containing these commands produces the output shown in Figure 7.11.

FIGURE 7.11   Revised column headings

```
                    Message indicates column
                    headings were cleared        New column
                                                 headings
columns cleared

Sales Rep                  Customer                              Current    Credit
Number/Name                Number/Name                           Balance     Limit
------------------------   -----------------------------------   --------  --------
  Available
    Credit
----------
20-Valerie Kaiser          148-Al's Appliance and Sport             6550      7500
       950

20-Valerie Kaiser          524-Kline's                             12762     15000
      2238

20-Valerie Kaiser          842-All Season                           8221      7500
      -721

35-Richard Hull            282-Brookings Direct                    431.5     10000
    9568.5

35-Richard Hull            408-The Everything Shop                5285.25     5000
    -285.25

35-Richard Hull            687-Lee's Sport and Appliance            2851      5000
      2149

35-Richard Hull            725-Deerfield's Four Seasons              248      7500
      7252

65-Juan Perez              356-Ferguson's                           5785      7500
      1715

65-Juan Perez              462-Bargains Galore                      3412     10000
      6588

65-Juan Perez              608-Johnson's Department Store           2106     10000
      7894

10 rows selected.
```

■ Changing Column Formats in a Report

You can use the COLUMN command to change more than just the column headings. You also can use it with a **FORMAT** clause to change the width of a column and the way the entries appear in a column. Example 10 illustrates these types of formatting changes.

: Change the format of the columns to allow the SLSR and CUST columns to display 20 and 30 characters, respectively. Display the data in the other columns with dollar signs and two decimal places.

The appropriate commands appear in Figure 7.12. The first COLUMN command changes the format of the SLSR column to A20 and the second changes the format of the CUST column to A30. The letter A indicates that the column is alphanumeric (another name for character); the number (20 or 30) indicates the column width in characters.

FIGURE 7.12   Formatting columns

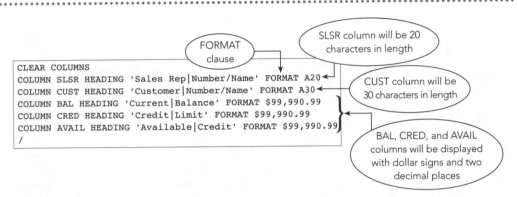

The next three COLUMN commands change the format of the three numeric columns in the view. In each case, the new format is $99,990.99. The 9s indicate that the value is numeric. The two 9s to the right of the decimal point indicate that each number will be displayed with two decimal places. The total number of 9s indicates the size of the column by representing the largest number that can be displayed. The dollar sign indicates that the values will be displayed as currency. Finally, the zero immediately to the left of the decimal point indicates that a value of zero will be displayed as $0.00. If you use all nines ($99,999.99), a value of zero will be blank.

One way to construct the appropriate format for the numeric columns is to write the largest number that the column can display. If the amount can be over $10,000 but must be less than $100,000, for example, the format will be $99,999.99. If the amount must be less than $1,000 but more than $100, the format will be $999.99. The final step is to determine whether a value of zero should be displayed. If you want a zero value to be displayed, change the nine immediately to the left of the decimal point to zero.

The result of running the script containing these commands appears in Figure 7.13. Notice the new format of the columns.

FIGURE 7.13   Report with formatting changes

```
columns cleared                                                     Rows no longer extend
                                                                    over multiple lines

Sales Rep            Customer                      Current      Credit     Available
Number/Name          Number/Name                   Balance      Limit         Credit
-------------------- ----------------------------- ----------- ----------- -----------
20-Valerie Kaiser    148-Al's Appliance and Sport   $6,550.00   $7,500.00     $950.00
20-Valerie Kaiser    524-Kline's                   $12,762.00  $15,000.00   $2,238.00
20-Valerie Kaiser    842-All Season                 $8,221.00   $7,500.00    -$721.00
35-Richard Hull      202-Brookings Direct             $431.50  $10,000.00   $9,568.50
35-Richard Hull      408-The Everything Shop        $5,285.25   $5,000.00    -$285.25
35-Richard Hull      687-Lee's Sport and Appliance  $2,851.00   $5,000.00   $2,149.00
35-Richard Hull      725-Deerfield's Four Seasons     $248.00   $7,500.00   $7,252.00
65-Juan Perez        356-Ferguson's                 $5,785.00   $7,500.00   $1,715.00
65-Juan Perez        462-Bargains Galore            $3,412.00  $10,000.00   $6,588.00
65-Juan Perez        608-Johnson's Department Store $2,106.00  $10,000.00   $7,894.00

10 rows selected.
```

New column widths

Numbers are displayed as currency

■ Adding a Title to a Report

The next step is to add a title to the report, as illustrated in Example 11.

EXAMPLE 11 :   Add a title that extends over two lines to the report. The first line is
"Customer Financial Report." The second line is "Organized by Sales Rep."

To add a title to the top of the report, use the **TTITLE** command, as shown in Figure 7.14. (To add a title at the bottom of the report, use the **BTITLE** command.) Then include the desired title within single quotation marks in the TTITLE command. To display the title on two lines, separate the lines with a vertical line, as shown in Figure 7.14.

FIGURE 7.14   Adding a title to the report

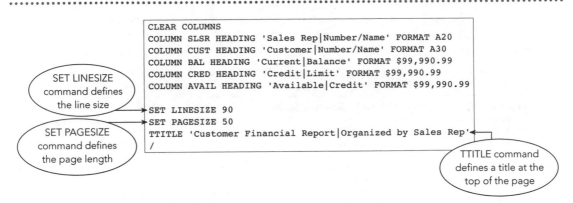

```
CLEAR COLUMNS
COLUMN SLSR HEADING 'Sales Rep|Number/Name' FORMAT A20
COLUMN CUST HEADING 'Customer|Number/Name' FORMAT A30
COLUMN BAL HEADING 'Current|Balance' FORMAT $99,990.99
COLUMN CRED HEADING 'Credit|Limit' FORMAT $99,990.99
COLUMN AVAIL HEADING 'Available|Credit' FORMAT $99,990.99
SET LINESIZE 90
SET PAGESIZE 50
TTITLE 'Customer Financial Report|Organized by Sales Rep'
/
```

SET LINESIZE command defines the line size

SET PAGESIZE command defines the page length

TTITLE command defines a title at the top of the page

When adding a title to a report, you might need to adjust the line size by using the **SET LINESIZE** command. The **line size**, which is the maximum number of characters each line can contain, determines where the title appears when it is centered across the line. In Figure 7.14, the SET LINESIZE command sets the line size to 90 characters. In this report, a line size of 90 characters is appropriate and places the title in the correct position on the line. In general, you can experiment with the line size to determine the best size for your report. The **page size**, which is the maximum number of lines per page, is specified by the **SET PAGESIZE** command.

The resulting report appears in Figure 7.15. Notice that the page number and date appear automatically with the title.

FIGURE 7.15   Title added at the top of the report

```
columns cleared                                                    Page number

Mon Aug 12      Current date                                       page     1
                              Customer Financial Report    Title appears
                                 Organized by Sales Rep     on two lines

Sales Rep             Customer                    Current      Credit   Available
Number/Name           Number/Name                 Balance       Limit      Credit
-------------------   -----------------------   ----------- ----------- -----------
20-Valerie Kaiser     148-Al's Appliance and Sport  $6,550.00   $7,500.00     $950.00
20-Valerie Kaiser     524-Kline's                  $12,762.00  $15,000.00   $2,238.00
20-Valerie Kaiser     842-All Season                $8,221.00   $7,500.00    -$721.00
35-Richard Hull       282-Brookings Direct            $431.50  $10,000.00   $9,568.50
35-Richard Hull       408-The Everything Shop        $5,285.25   $5,000.00    -$285.25
35-Richard Hull       687-Lee's Sport and Appliance  $2,851.00   $5,000.00   $2,149.00
35-Richard Hull       725-Deerfield's Four Seasons     $248.00   $7,500.00   $7,252.00
65-Juan Perez         356-Ferguson's                 $5,785.00   $7,500.00   $1,715.00
65-Juan Perez         462-Bargains Galore            $3,412.00  $10,000.00   $6,588.00
65-Juan Perez         608-Johnson's Department Store $2,106.00  $10,000.00   $7,894.00

10 rows selected.
```

■ Grouping Data in a Report

Just as you can group data by using SQL queries, you can also group data in reports by using the **BREAK** command. You use the BREAK command to identify a column (or collection of columns) on which to group the data. The value in the column is displayed only at the beginning of the group. In addition, you can specify a number of lines to skip after each group. Example 12 illustrates the use of the BREAK command in grouping data. The example also removes the message "10 rows selected." from the end of the report.

EXAMPLE 12 : Group the rows in the report by the SLSR column. In addition, remove the message at the end of the report that indicates the number of rows selected.

To group rows by the SLSR column, the command is BREAK ON REPORT ON SLSR, as shown in Figure 7.16. For the BREAK command to work properly, you need to sort the data on the indicated column. In this example, the data is sorted correctly because the SELECT command that produced the data included an ORDER BY SLSR clause.

FIGURE 7.16   Adding a break

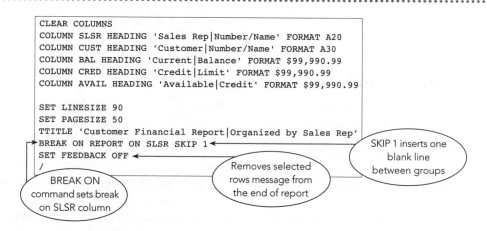

```
CLEAR COLUMNS
COLUMN SLSR HEADING 'Sales Rep|Number/Name' FORMAT A20
COLUMN CUST HEADING 'Customer|Number/Name' FORMAT A30
COLUMN BAL HEADING 'Current|Balance' FORMAT $99,990.99
COLUMN CRED HEADING 'Credit|Limit' FORMAT $99,990.99
COLUMN AVAIL HEADING 'Available|Credit' FORMAT $99,990.99

SET LINESIZE 90
SET PAGESIZE 50
TTITLE 'Customer Financial Report|Organized by Sales Rep'
BREAK ON REPORT ON SLSR SKIP 1
SET FEEDBACK OFF
/
```

SKIP 1 inserts one blank line between groups

Removes selected rows message from the end of report

BREAK ON command sets break on SLSR column

The 1 in the **SKIP** clause at the end of the BREAK command inserts one blank line between groups. The **SET FEEDBACK OFF** command turns off the message indicating the number of rows selected by the query.

The result of executing these commands appears in Figure 7.17. Notice that the rows are grouped by sales rep, with the sales rep number and name appearing only once. A blank line separates the groups, and the message indicating the number of rows selected is not displayed.

FIGURE 7.17   Blank lines added to the report

```
columns cleared

Mon Aug 12                                                        page    1
                           Customer Financial Report
                             Organized by Sales Rep

Sales Rep            Customer                   Current     Credit   Available
Number/Name          Number/Name                Balance      Limit      Credit
-------------------  ---------------------------  ----------  ----------  ----------
20-Valerie Kaiser    148-Al's Appliance and Sport  $6,550.00  $7,500.00     $950.00
                     524-Kline's                 $12,762.00 $15,000.00   $2,238.00
                     842-All Season               $8,221.00  $7,500.00    -$721.00

35-Richard Hull      282-Brookings Direct           $431.50 $10,000.00   $9,568.50
                     408-The Everything Shop      $5,285.25  $5,000.00    -$285.25
                     687-Lee's Sport and Appliance $2,851.00  $5,000.00   $2,149.00
                     725-Deerfield's Four Seasons   $248.00  $7,500.00   $7,252.00

65-Juan Perez        356-Ferguson's               $5,785.00  $7,500.00   $1,715.00
                     462-Bargains Galore          $3,412.00 $10,000.00   $6,588.00
                     608-Johnson's Department Store $2,106.00 $10,000.00  $7,894.00
```

Line breaks

No feedback appears

Including Totals and Subtotals in a Report

A total that appears after each group is called a **subtotal**. To calculate a subtotal, you must include a BREAK command to group the rows. Then you can use a **COMPUTE** command to indicate the computation for the subtotal, as shown in Example 13.

EXAMPLE 13 : Include totals and subtotals in the report for the BAL and AVAIL columns.

The COMPUTE command uses statistical functions to calculate the values to include in the report. The SQL statistical functions are shown in Figure 7.18.

FIGURE 7.18

Reporting command summary

Statistical function	Result of calculation
AVG	Average of values in a column
COUNT	Number of rows in a table
MAX	Largest value in a column
MIN	Smallest value in a column
STDEV	Standard deviation of values in a column
SUM	Sum of values in a column
VARIANCE	Variance of values in a column

In Example 13, SUM is the appropriate function to use. The script shown in Figure 7.19 shows the four COMPUTE commands needed to include totals and subtotals in the report.

FIGURE 7.19   Adding computations to a report

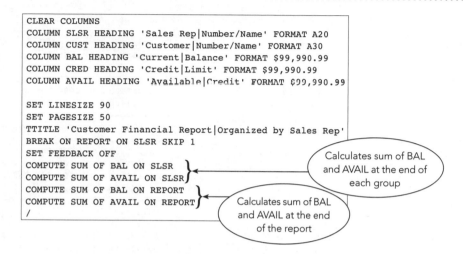

```
CLEAR COLUMNS
COLUMN SLSR HEADING 'Sales Rep|Number/Name' FORMAT A20
COLUMN CUST HEADING 'Customer|Number/Name' FORMAT A30
COLUMN BAL HEADING 'Current|Balance' FORMAT $99,990.99
COLUMN CRED HEADING 'Credit|Limit' FORMAT $99,990.99
COLUMN AVAIL HEADING 'Available|Credit' FORMAT $99,990.99

SET LINESIZE 90
SET PAGESIZE 50
TTITLE 'Customer Financial Report|Organized by Sales Rep'
BREAK ON REPORT ON SLSR SKIP 1
SET FEEDBACK OFF
COMPUTE SUM OF BAL ON SLSR
COMPUTE SUM OF AVAIL ON SLSR
COMPUTE SUM OF BAL ON REPORT
COMPUTE SUM OF AVAIL ON REPORT
/
```

Calculates sum of BAL and AVAIL at the end of each group

Calculates sum of BAL and AVAIL at the end of the report

Notice that the commands in Figure 7.19 contain ON clauses that include the desired computations and the column names on which the computations are to occur. The **ON** clause indicates the point at which the computation is to occur; it must contain the column name (or the word REPORT) that is included in the BREAK command.

The computations that end with ON REPORT are displayed only once at the end of the report. In this report, the computations represent grand totals. The computations associated with the other breaks occur at the end of the indicated groups. In this report, the computations represent subtotals that are displayed after the group of customers of a particular sales rep. The result is shown in Figure 7.20.

FIGURE 7.20 Totals and subtotals included

```
columns cleared

Mon Aug 12                                                            page    1
            Break (end of records   Customer Financial Report
            for sales rep 20)          Organized by Sales Rep


Sales Rep           Customer                    Current      Credit    Available
Number/Name         Number/Name                 Balance       Limit       Credit
-----------------   -----------------------   ----------   ----------  ----------
20-Valerie Kaiser   148-Al's Appliance and Sport  $6,550.00   $7,500.00     $950.00
                    524-Kline's                  $12,762.00  $15,000.00   $2,238.00
                    842-All Season                $8,221.00   $7,500.00    -$721.00
*******************                           Subtotals for ----------  ----------
sum                                           BAL and AVAIL $27,533.00         $2,467.00

35-Richard Hull     282-Brookings Direct           $431.50  $10,000.00   $9,568.50
                    408-The Everything Shop      $5,285.25   $5,000.00    -$285.25
                    687-Lee's Sport and Appliance $2,851.00   $5,000.00   $2,149.00
                    725-Deerfield's Four Seasons    $248.00   $7,500.00   $7,252.00
*******************                              ----------             ----------
sum                                              $8,815.75              $18,684.25

65-Juan Perez       356-Ferguson's               $5,785.00   $7,500.00   $1,715.00
                    462-Bargains Galore          $3,412.00  $10,000.00   $6,588.00
                    608-Johnson's Department Store $2,106.00 $10,000.00   $7,894.00
*******************                              ----------             ----------
sum                                             $11,303.00              $16,197.00
                        Grand totals for
                        BAL and AVAIL
                                                ----------             ----------
sum                                             $47,651.75             $37,348.25
```

If the report is to be displayed on the screen, you will want to set the page size to an appropriate value for the screen, such as 24. In addition, you can include the **SET PAUSE ON** command, which causes Oracle to pause after each screen of data. To see the next screen, press the Enter key.

Note: In some SQL implementations, the SET PAUSE ON command causes the report to pause when you run the script. Press the Enter key to see the first and subsequent pages.

The report includes subtotals to indicate the balances and available credit limits for every customer of each sales rep. Subtotals represent a subset of the overall total. Grand totals of the balance and available credit amounts for all customers appear at the end of the report.

■ Sending the Report to a File

In many cases, viewing the results of a query on the screen is sufficient. In other cases, you might want to print a copy; this is especially true for reports. The exact manner in which you print a report depends on the DBMS. If you have any questions about printing, consult your DBMS documentation.

To print a report using Oracle, you first send the output of the query to a file by using the **SPOOL** command. (The process of sending printed output to a file rather than directly to a printer is called **spooling**, and this is where the command gets its name.) After spooling output to a file, you can print the contents of the file just as you would print the contents of any other file.

EXAMPLE 14 : Send the report created in the previous examples to a file named
                : SLSR_REPORT_OUTPUT.SQL.

The SPOOL SLSR_REPORT_OUTPUT.SQL command shown in Figure 7.21 sends the output of subsequent commands to the file named SLSR_REPORT_OUTPUT.SQL. The final command (**SPOOL OFF**) turns off spooling and stops any further output from being sent to the SLSR_REPORT_OUTPUT.SQL file.

FIGURE 7.21
. . . .
Sending the
report to a file

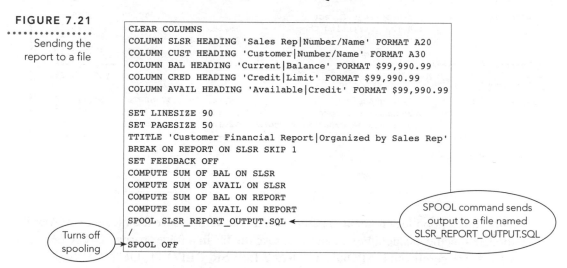

```
CLEAR COLUMNS
COLUMN SLSR HEADING 'Sales Rep|Number/Name' FORMAT A20
COLUMN CUST HEADING 'Customer|Number/Name' FORMAT A30
COLUMN BAL HEADING 'Current|Balance' FORMAT $99,990.99
COLUMN CRED HEADING 'Credit|Limit' FORMAT $99,990.99
COLUMN AVAIL HEADING 'Available|Credit' FORMAT $99,990.99

SET LINESIZE 90
SET PAGESIZE 50
TTITLE 'Customer Financial Report|Organized by Sales Rep'
BREAK ON REPORT ON SLSR SKIP 1
SET FEEDBACK OFF
COMPUTE SUM OF BAL ON SLSR
COMPUTE SUM OF AVAIL ON SLSR
COMPUTE SUM OF BAL ON REPORT
COMPUTE SUM OF AVAIL ON REPORT
SPOOL SLSR_REPORT_OUTPUT.SQL
/
SPOOL OFF
```

SPOOL command sends output to a file named SLSR_REPORT_OUTPUT.SQL

Turns off spooling

When you run the new script, the report is displayed again on the screen. As it is displayed on the screen (see Figure 7.22), it also is being written to the file. After the spooling process is complete, the report is stored in the SLSR_REPORT_OUTPUT.SQL file. You can print the file, edit it, include it in a document, or use it as needed in other ways.

FIGURE 7.22 Running the final report

```
columns cleared                                      ⟨Report also sent⟩
                                                     ⟨to spool file⟩

Mon Aug 12                                                    page    1
                          Customer Financial Report
                          Organized by Sales Rep

Sales Rep              Customer                  Current      Credit    Available
Number/Name            Number/Name               Balance      Limit        Credit
--------------------   ------------------------  -----------  ----------- -----------
20-Valerie Kaiser      148-Al's Appliance and Sport  $6,550.00  $7,500.00    $950.00
                       524-Kline's                  $12,762.00 $15,000.00  $2,238.00
                       842-All Season                $8,221.00  $7,500.00   -$721.00
********************                              -----------              -----------
sum                                                 $27,533.00              $2,467.00


35-Richard Hull        282-Brookings Direct           $431.50 $10,000.00  $9,568.50
                       408-The Everything Shop       $5,285.25  $5,000.00   -$285.25
                       687-Lee's Sport and Appliance $2,851.00  $5,000.00  $2,149.00
                       725-Deerfield's Four Seasons   $248.00   $7,500.00  $7,252.00
********************                              -----------              -----------
sum                                                  $8,815.75             $18,684.25


65-Juan Perez          356-Ferguson's               $5,785.00  $7,500.00  $1,715.00
                       462-Bargains Galore          $3,412.00 $10,000.00  $6,588.00
                       608-Johnson's Department Store $2,106.00 $10,000.00 $7,894.00
********************                              -----------              -----------
sum                                                 $11,303.00             $16,197.00


                                                 -----------              -----------
sum                                                 $47,651.75             $37,348.25
```

Note: It is a good idea to include pathnames if you want to save the file on drive A or in another specific location. For example, to save the file in a folder named ORACLE on drive A, the command is SPOOL A:\ORACLE\SLSR_REPORT_OUTPUT.SQL.

■ Completing the Script to Produce the Report

You can include the additional commands shown in Figure 7.23 in the script to finish it.

FIGURE 7.23   Completed script

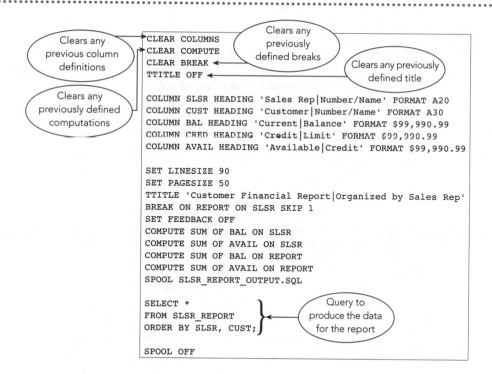

The first command, CLEAR COLUMNS, clears any previous column definitions. The next two commands have the same purpose. The **CLEAR COMPUTE** command clears any previously specified computations, and the **CLEAR BREAK** command clears any previous breaks. The **TTITLE OFF** command turns off any previously specified title at the top of the report (the **BTITLE OFF** command would turn off any previously specified title at the bottom of the report).

Notice that the actual SQL query for the report is included in the completed script. Without this query, a user who runs this script would have to execute the query first and then run the script. The script won't work without executing the query, so it is a good idea to include the query in the script to avoid this problem.

The remaining commands in the script are the same ones you used earlier. Running the script shown in Figure 7.23 spools the report shown in Figure 7.24 to a file named SLSR_REPORT_OUTPUT.SQL. Notice that the query appears at the beginning of the spooled file. You can remove the query from the file if desired by opening the file in a text editor such as Notepad.

FIGURE 7.24   Completed report

```
SELECT *
FROM SLSR_REPORT
ORDER BY SLSR, CUST;        Query appears
                           first in spool file
Mon Aug 12                                                              page    1
                           Customer Financial Report
                           Organized by Sales Rep

Sales Rep                Customer                    Current     Credit   Available
Number/Name              Number/Name                 Balance      Limit     Credit
--------------------     ---------------------------  ----------- ----------- -----------
20-Valerie Kaiser        148-Al's Appliance and Sport  $6,550.00  $7,500.00    $950.00
                         524-Kline's                  $12,762.00 $15,000.00  $2,238.00
                         842-All Season                $8,221.00  $7,500.00   -$721.00
*******************                                   ----------            -----------
sum                                                  $27,533.00             $2,467.00

35-Richard Hull          282-Brookings Direct           $431.50 $10,000.00  $9,568.50
                         408-The Everything Shop       $5,285.25  $5,000.00  -$285.25
                         687-Lee's Sport and Appliance $2,851.00  $5,000.00  $2,149.00
                         725-Deerfield's Four Seasons    $248.00  $7,500.00  $7,252.00
*******************                                   ----------            -----------
sum                                                   $8,815.75            $18,684.25

65-Juan Perez            356-Ferguson's                $5,785.00  $7,500.00  $1,715.00
                         462-Bargains Galore           $3,412.00 $10,000.00  $6,588.00
                         608-Johnson's Department Store $2,106.00 $10,000.00  $7,894.00
*******************                                   ----------            -----------
sum                                                  $11,303.00            $16,197.00


                                                     ----------            -----------
sum                                                  $47,651.75            $37,348.25
```

In this chapter, you learned how to use several functions and to perform calculations with dates. You learned how to concatenate columns. You also learned about several important commands in Oracle that are used to create and format reports. Figure 7.25 lists these report-formatting commands and their descriptions.

FIGURE 7.25   Reporting command summary

Command	Description
BREAK	Groups data in a report on a specified column
BTITLE	Adds a title at the bottom of a report
BTITLE OFF	Clears any previously specified title at the bottom of a report
CLEAR BREAK	Clears any previously specified report breaks
CLEAR COLUMNS	Clears any previous column changes
CLEAR COMPUTE	Clears any previously specified report computations
COLUMN	Changes the name of a column in a report
COMPUTE	Calculates a count, minimum, maximum, sum, average, standard deviation, or variance on the values in a column in a report
HEADING	Assigns a new heading to a column in a report
ON REPORT	Indicates that a calculation is to be performed on all values in the report
RTRIM	Deletes extra spaces that appear after a value in a column
SET FEEDBACK OFF	Turns off the message indicating the number of rows selected by a query in a report
SET LINESIZE	Indicates the maximum number of characters on a line
SET PAGESIZE	Indicates the number of lines on a page
SET PAUSE ON	Indicates that the screen display should pause after each screen of data
SKIP	Inserts the defined number of blank lines between groups in a report
SPOOL	Sends output to a file for printing or editing
SPOOL OFF	Stops spooling to a file
TTITLE	Adds a title at the top of a report
TTITLE OFF	Clears any previously specified title at the top of a report

■ SUMMARY

- There are functions whose results are based on the values in single records. UPPER and LOWER are two examples of functions that act on character data. UPPER converts each lowercase letter in the argument to uppercase. LOWER converts each uppercase letter in the argument to lowercase.

- ROUND and FLOOR are two examples of functions that act on numeric data. ROUND produces its result by rounding the value to the specified number of decimal places. FLOOR produces its result by truncating (removing) everything to the right of the decimal point.

- The ADD_MONTHS function in Oracle adds a specific number of months to a date. In MySQL, use the ADDDATE function for the same purpose; in Access, use the DATEADD function.

- In Oracle you can add a specific number of days to a date using normal addition. In MySQL, use the ADDDATE function for this purpose; in Access, use the DATEADD function.

- You can subtract one date from another to produce the number of days between two dates.

- To obtain today's date, use SYSDATE in Oracle, CURDATE() in MySQL, and DATE() in Access.

- To concatenate values in character columns, separate the column names with two vertical lines (||). Use the RTRIM function to delete any extra spaces that follow the values. In Access, use the ampersand (&) to concatenate values. In MySQL, use the CONCAT function.

- You can produce script files to create views and format reports.

- The data for a report can come from a table or a view.

- Use the COLUMN command to change a column heading. Use the HEADING clause to assign a new heading name. Type a single vertical line (|) to break a column heading over two lines.

- Use the CLEAR COLUMNS command to clear any previously defined column specifications.

- Use the COLUMN command with a FORMAT clause to change the format of column values.

- Use the TTITLE or BTITLE command to add a title at the top or bottom of a report, respectively.

- Use the SET LINESIZE command to set the number of characters per line. Use the SET PAGESIZE command to set the number of lines per page.

- Use the BREAK command to group data in a report.

- Use the SKIP command to insert blank lines between groups in a report.

- Use the SET FEEDBACK OFF command to turn off the message indicating the number of records selected in the query results.

- Use the BREAK and COMPUTE commands and an appropriate statistical function to calculate data in a report, such as totals and subtotals.

- Use the SET PAUSE ON command to pause a report after each screen of data or to change the number of report lines to display.

- Use the SPOOL command to send a report to a file for printing or editing. Use the SPOOL OFF command to stop spooling.

- Use the CLEAR COMPUTE, CLEAR BREAK, and TTITLE OFF commands to clear any previously defined computations, breaks, or top titles, respectively.

■ KEY TERMS

ADDDATE (MySQL)
ADD_MONTHS (Oracle)
argument
BREAK
BTITLE
BTITLE OFF
CLEAR BREAK
CLEAR COLUMNS
CLEAR COMPUTE
COLUMN
COMPUTE
concatenate
concatenation
CURDATE (MySQL)
DATE (Access)
DATEADD (Access)
FLOOR (Oracle)
FORMAT
HEADING
line size

LCASE (Access)
LOWER (Oracle)
ON
page size
ROUND (Oracle)
RTRIM
SET FEEDBACK OFF
SET LINESIZE
SET PAGESIZE
SET PAUSE ON
SKIP
SPOOL
SPOOL OFF
spooling
subtotal
SYSDATE (Oracle)
TTITLE
TTITLE OFF
UCASE (Access)
UPPER (Oracle)

■ REVIEW QUESTIONS

1. How do you convert letters to uppercase? How do you convert letters to lowercase?

2. How do you round a number to a specific number of decimal places? How do you remove everything to the right of the decimal place?

3. How do you add months to a date? How do you add days to a date? How would you find the number of days between two dates?

4. How do you obtain today's date?

5. How do you concatenate values in character columns?

6. Which function deletes extra spaces at the end of a value?

7. How do you select the data to use in a report?

8. Which command and clause changes a column heading?

9. How can you print a column heading on two lines?

10. Which command and clause changes the format of the values in a column in a report?

11. Which command adds a title at the top of a page? Which command adds a title at the bottom of a page?

12. Which command sets the number of characters per line in a report?

13. Which command sets the number of lines per page in a report?

14. Which command groups data in a report?

15. Which command calculates statistics, such as totals and subtotals, in a report?

16. Which command pauses a report after each screen of data?

17. Which command sends a report to a file?

■EXERCISES (Premiere Products)

Use SQL and the Premiere Products database to complete the following exercises.

1. List the part number and description for all parts. The part descriptions should appear in uppercase letters.

2. List the customer number and name for all customers located in the city of Grove. Your query should ignore case. For example, a customer with the city Grove should be included as should customers whose city is GROVE, grove, GrOvE, and so on.

3. List the customer number, name, and balance for all customers. The balance should be rounded to the nearest dollar.

4. Premiere Products is running a promotion that is valid for up to 20 days after an order is placed. List the order number, customer number, customer name, and the promotion date for each order. The promotion date is the date that is exactly 20 days after the order was placed.

5. If you are using Oracle, produce the report shown in Figure 7.26. Create a view for the report, if necessary. Write the script to produce the report and make any changes to LINESIZE and/or PAGE-SIZE that you feel are necessary.

FIGURE 7.26

```
CUSTOMER                               CUSTOMER          CUSTOMER
NAME                                   ADDRESS           CITY/STATE/ZIP
-------------------------------------- ---------------   ------------------------
Al's Appliance and Sport               2837              Fillmore, FL 33336
                                       Greenway

Brookings Direct                       3827              Grove, FL 33321
                                       Devon

Ferguson's                             382               Northfield, FL 33146
                                       Wildwood

The Everything Shop                    1828              Crystal, FL 33503
                                       Raven

Bargains Galore                        3829              Grove, FL 33321
                                       Central

Kline's                                838               Fillmore, FL 33336
                                       Ridgeland

Johnson's Department Store             372               Sheldon, FL 33553
                                       Oxford

Lee's Sport and Appliance              282               Altonville, FL 32543
                                       Evergreen

Deerfield's Four Seasons               282               Sheldon, FL 33553
                                       Columbia

All Season                             28 Lakeview       Grove, FL 33321

10 rows selected.
```

6. If you are using Oracle, produce the report shown in Figure 7.27. Create a view for the report, if necessary. Write the script to produce the report and make any changes to LINESIZE and/or PAGESIZE that you feel are necessary. Make sure that the commission rates appear with zeroes to the left of the decimal points.

7. If you are using Oracle, produce the report shown in Figure 7.28. Create a view for the report, if necessary. Write the script to produce the report and make any changes to LINESIZE and/or PAGESIZE that you feel are necessary.

FIGURE 7.27

```
Tue Aug 20                                                    page     1
                                 Sales Rep Commissions
                                       and Rates

     Sales Rep                       Total
     Name                          Commission      RATE
     ------------------------------ ----------- -----------
     Valerie Kaiser                 $20,542.50      0.05
     Richard Hull                   $39,216.00      0.07
     Juan Perez                     $23,487.00      0.05
```

FIGURE 7.28

```
Tue Aug 20                                                    page     1
                              Parts Ordered Report

     Part Number/         Order     Units     Quoted     Actual
     Description          Number   Ordered    Price      Price  Difference
     -------------------- -------- ------- ---------- ---------- ----------
     AT94-Iron            21608        11    $21.95     $24.95     $3.00
     ********************          -------
     sum                              11


     BV06-Home Gym        21617         2   $794.95    $794.95     $0.00
     ********************          -------
     sum                               2


     CD52-Microwave Oven  21617         4   $150.00    $165.00    $15.00
     ********************          -------
     sum                               4


     DR93-Gas Range       21610         1   $495.00    $495.00     $0.00
                          21619         1   $495.00    $495.00     $0.00
     ********************          -------
     sum                               2


     DW11-Washer          21610         1   $399.99    $399.99     $0.00
     ********************          -------
     sum                               1


     KL62-Dryer           21613         4   $329.95    $349.95    $20.00
     ********************          -------
     sum                               4


     KT03-Dishwasher      21614         2   $595.00    $595.00     $0.00
     ********************          -------
     sum                               2


     KV29-Treadmill       21623         2 $1,290.00  $1,390.00   $100.00
     ********************          -------
     sum                               2


                                   -------
     sum                              28
```

■EXERCISES (Henry Books)

Use SQL and the Henry Books database to complete the following exercises.

1. List the author number, first name, and last name for all authors. The first name should appear in lowercase letters and the last name should appear in uppercase letters.

2. List the publisher code and name for all publishers located in the city of New York. Your query should ignore case. For example, a customer with the city New York should be included as should customers whose city is NEW YORK, New york, NeW yOrK, and so on.

3. List the book code, title, and price for all books. The price should be rounded to the nearest dollar.

4. If you are using Oracle, produce the report shown in Figure 7.29. (*Note:* Your page number might appear in a different position.) Create a view for the report, if necessary. Write the script to produce the report and make any changes to LINE-SIZE and/or PAGESIZE that you feel are necessary.

FIGURE 7.29

```
Wed Aug 21                                                        page    1
                              Inventory List
                              Henry Books


  Br                                                              Units
Numb  Book Title                          Publisher Name          Price On Hand
----  ----------------------------------  ------------------------------ -------- -------
   1  A Deepness in the Sky               TB-Tor Books                    $7.19      2
      The Stranger                        VB-Vintage Books                $8.00      1
      Harry Potter and the Prisoner of Azkaban ST-Scholastic Trade      $13.96      3
      Travels with Charley                PE-Penguin USA                  $7.95      1
      Group: Six People in Search of a Life  BP-Berkley Publishing       $10.40      2
      Electric Light                      FS-Farrar Straus and Giroux    $14.00      3
      A Guide to SQL                      CT-Course Technology           $37.95      1
      Black House                         RH-Random House                $18.81      2
      To Kill a Mockingbird               HC-HarperCollins Publishers    $18.00      2
      The Grapes of Wrath                 PE-Penguin USA                 $13.00      2
      When Rabbit Howls                   JP-Jove Publications            $6.29      3
****                                                                            -------
sum                                                                                22


   2  Magic Terror                        FA-Fawcett Books                $7.99      2
      Second Wind                         PU-Putnam Publishing Group     $24.95      1
      The Stranger                        VB-Vintage Books                $8.00      3
      Dreamcatcher: A Novel               SC-Scribner                    $19.60      4
      Treasure Chests                     TA-Taunton Press               $24.46      1
      Beloved                             PL-Plume                       $12.95      3
      The Catcher in the Rye              LB-Lb Books                     $5.99      3
      The Grapes of Wrath                 PE-Penguin USA                 $13.00      1
      The Fall                            VB-Vintage Books                $8.00      2
      Franny and Zooey                    LB-Lb Books                     $5.99      2
      Band of Brothers                    TO-Touchstone Books             $9.60      2
      Jazz                                PL-Plume                       $12.95      4
      The Soul of a New Machine           BY-Back Bay Books              $11.16      1
      Nine Stories                        LB-Lb Books                     $5.99      1
      The Edge                            JP-Jove Publications            $6.99      1
****                                                                            -------
sum                                                                                31

   3  Venice                              SS-Simon and Schuster          $24.50      2
      Second Wind                         PU-Putnam Publishing Group     $24.95      2
      Slay Ride                           JP-Jove Publications            $6.99      3
      The Grapes of Wrath                 PE-Penguin USA                 $13.00      3
      Song of Solomon                     PL-Plume                       $14.00      5
      Godel, Escher, Bach                 BA-Basic Books                 $14.00      1
      East of Eden                        PE-Penguin USA                 $12.95      2
      Jazz                                PL-Plume                       $12.95      3
      The Soul of a New Machine           BY-Back Bay Books              $11.16      2
      Harry Potter and the Prisoner of Azkaban ST-Scholastic Trade      $13.96      2
```

FIGURE 7.29 (continued)

```
Wed Aug 21                                                              page    2
                                  Inventory List
                                  Henry Books

  Br                                                                       Units
Numb Book Title                          Publisher Name            Price On Hand
---- ------------------------------------ ------------------------------ -------- -------
   3 Of Mice and Men                      PE-Penguin USA             $6.95       2
     Dreamcatcher: A Novel                SC-Scribner               $19.60       2
****                                                                       -------
sum                                                                            29

   4 Second Wind                          PU-Putnam Publishing Group $24.95       3
     Harry Potter and the Prisoner of Azkaban ST-Scholastic Trade  $13.96       1
     Van Gogh and Gauguin                 WP-Westview Press         $21.00       3
     The Catcher in the Rye               LB-Lb Books                $5.99       2
     The Grapes of Wrath                  PE-Penguin USA            $13.00       2
     Song of Solomon                      PL-Plume                  $14.00       2
     Harry Potter and the Goblet of Fire  ST-Scholastic Trade       $18.16       1
     Catch-22                             SC-Scribner               $12.00       2
     Electric Light                       FS-Farrar Straus and Giroux $14.00     1
****                                                                       -------
sum                                                                            17

                                                                           -------
sum                                                                            99
```

▇EXERCISES (Alexamara Marina Group)

Use SQL and the Alexamara Marina Group database to complete the following exercises.

1. List the owner number, first name, and last name for all owners. The first name should appear in uppercase letters and the last name should appear in lowercase letters.

2. List the owner number and last name for all owners located in the city of Bowton. Your query should ignore case. For example, a customer with the city Bowton should be included as should customers whose city is BOWTON, BowTon, BoWtOn, and so on.

3. Alexamara is offering a discount for owners who sign up early for slips for next year. The discount is 1.75% of the rental fee. For each slip, list the marina number, slip number, owner number, owner's last name, rental fee, and discount. The discount should be rounded to the nearest dollar.

4. Alexamara tries to complete each service request by the next service date, but sometimes this is not possible. Alexamara has a policy, however, to address all service requests within 14 days of the next service date. That is, the last possible date to complete each service request would be 14 days after the next service date. Produce a report that includes the service ID, marina number, slip number, owner number, owner name, category description, service description, next service date, and last possible date for each service request.

5. If you are using Oracle, produce the report shown in Figure 7.30. (*Note:* Your page number might appear in a different position.) Create a view for the report, if necessary. Write the script to produce the report and make any changes to LINESIZE and/or PAGESIZE that you feel are necessary.

FIGURE 7.30

```
Tue Aug 20                                                               page     1
                                     Marina Slip List
                                   Alexamara Marina Group

Marina                  Slip   Boat                     Owner                    Rental
Name                    Number Name                     Name                       Fee
--------------------    ------ ------------------------ ------------------------ ------------
Alexamara Central       1      Bravo                    Bruce and Jean Adney      $1,800.00
                        4      Mermaid                  Mary Blake                $2,500.00
                        2      Chinook                  Daniel Feenstra           $1,800.00
                        5      Axxon II                 Peter Norton              $4,200.00
                        3      Listy                    Becky and Dave Smeltz     $2,000.00
                        6      Karvel                   Ashton Trent              $4,200.00
********************                                                             ------------
sum                                                                              $16,500.00

Alexamara East          A1     Anderson II              Bill Anderson             $3,800.00
                        B2     Anderson III             Bill Anderson             $2,600.00
                        A2     Our Toy                  Sandy and Bill Elend      $3,800.00
                        B1     Gypsy                    Maria Juarez              $2,400.00
                        A3     Escape                   Alyssa Kelly              $3,600.00
********************                                                             ------------
sum                                                                              $16,200.00

                                                                                ------------
sum                                                                              $32,700.00
```

CHAPTER 8

Embedded SQL

OBJECTIVES

- Embed SQL commands in PL/SQL programs

- Retrieve single rows using embedded SQL

- Update a table using embedded INSERT, UPDATE, and DELETE commands

- Use cursors to retrieve multiple rows in embedded SQL

- Update a database using cursors

- Manage errors in programs containing embedded SQL commands

- Use SQL in a language that does not support embedded SQL commands

Introduction

SQL is a powerful non-procedural language in which you communicate tasks to the computer using simple commands. As in other non-procedural languages, you can accomplish many tasks using a single, relatively simple command. A procedural language, on the other hand, is one in which you must give the computer the step-by-step process for accomplishing tasks, which might require many lines of code. PL/SQL, which was developed by Oracle as an extension of SQL, is an example of a procedural language. This chapter uses PL/SQL to illustrate how to **embed** SQL commands in another language, that is, how to include SQL commands directly within programs written in another language.

Although SQL and other non-procedural languages are well-equipped to store and query data, sometimes you might need to complete tasks that are beyond the capabilities of SQL. In such cases, you need to use a procedural language. Fortunately, you can embed SQL commands in a procedural language to capitalize on the advantages of SQL. You can use SQL for some tasks, and then include embedded SQL commands in the procedural language to accomplish tasks that are beyond the capabilities of SQL.

In this chapter, you will learn how to embed SQL commands in PL/SQL. The process of embedding SQL in other programming languages, such as C or COBOL, is very similar. The examples in this chapter illustrate how to use embedded SQL commands to retrieve a single row, insert new rows, and update and delete existing rows. Finally, you will learn how to retrieve multiple rows. Executing a SELECT command that retrieves more than one row presents a problem for a language such as PL/SQL, which is oriented toward processing one record at a time.

The final portion of this chapter deals with Microsoft Access. You cannot simply embed SQL commands in Access programs the way you can in PL/SQL. There are ways to use the commands, however, as you'll learn in the final section.

Note: This chapter assumes that you have some programming background and does not cover programming basics. To understand the first part of this chapter, you should be familiar with variables, declaring variables, and creating procedural code, including IF statements and loops. To understand the Access section at the end of the chapter, you should be familiar with Function and Sub Procedures, and the process for sequentially accessing all records in a recordset, such as using a loop to process all the records in a table.

ORACLE USER : The PL/SQL programs in this chapter are interactive. As such, you must run them in SQL*Plus, which permits input during program execution, rather than in SQL*Plus Worksheet, which does not. You can write the programs and save them as script files in either SQL*Plus or SQL*Plus

Worksheet. If you use SQL*Plus Worksheet to write your scripts, you will need to have SQL*Plus running simultaneously to run them. You can use the Windows taskbar to switch between programs.

ACCESS USER If you are using Access, you will not be able to complete the material concerning PL/SQL programs, but it is very important that you read it so that you will understand these important concepts. You will, however, be able to complete the steps in the "Using SQL in Microsoft Access Programs" section.

MYSQL USER If you are using MySQL, you will not be able to complete the material in this chapter, but it is very important that you read the chapter so that you will understand these important concepts.

■ Using Prompt Variables

There are many sources of input for SQL commands. The input can come from an on-screen form or be passed as arguments from some procedure. In the PL/SQL commands in this text, the input will come from **prompt variables** for which the user is prompted to enter a value when the program is run. You can use prompt variables in PL/SQL programs and in individual SQL commands, as illustrated in Example 1. To designate a variable as a prompt variable, precede the variable name with an ampersand (&).

EXAMPLE 1 List the last name of the sales rep whose number is contained in the prompt variable I_REP_NUM.

The SQL command and its execution are shown in Figure 8.1. The WHERE clause contains the prompt variable I_REP_NUM, which is enclosed in single quotation marks because it is a character column. When the program is executed, the prompt asks the user to enter a rep number. In Figure 8.1, the user entered the value 20, which was inserted in the command and automatically enclosed in single quotation marks (WHERE REP_NUM = '20'). (If the command did not include these single quotation marks, the user would need to enter the value as '20'.) The next two lines display the old (original) line in the command and the new line after the user enters a value. In the new line, the value '20' has replaced the '&I_REP_NUM' prompt variable.

FIGURE 8.1
· · · · · · · · · · · · · ·
SELECT
command with a
prompt variable

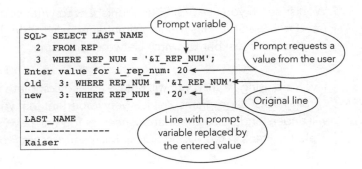

SQL> SELECT LAST_NAME
  2   FROM REP
  3   WHERE REP_NUM = '&I_REP_NUM';
Enter value for i_rep_num: 20
old   3: WHERE REP_NUM = '&I_REP_NUM'
new   3: WHERE REP_NUM = '20'

LAST_NAME

Kaiser

Note: If a prompt variable corresponds to a numeric field, you do not need to enclose it in single quotation marks.

■ PL/SQL Programs

You can embed SQL commands in PL/SQL programs as illustrated in the following examples. You create and save the programs as script files. To run the programs, run the script files.

Retrieving a Single Row and Column

Example 2 illustrates using embedded SQL to retrieve a single row and column from a table.

EXAMPLE 2 ⋮ Write a PL/SQL program to obtain the last name of the sales rep whose number is contained in the prompt variable I_REP_NUM, place it in the variable I_LAST_NAME, and then display the contents of I_LAST_NAME.

To place the results of a command in a variable, use the INTO clause. For this example, the SQL command would be:

```
SELECT LAST_NAME
INTO I_LAST_NAME
FROM REP
WHERE REP_NUM = '&I_REP_NUM;'
```

When this command is executed, Oracle will prompt the user for a value for I_REP_NUM, use that value to retrieve the last name of the sales rep whose number is equal to the number the user entered, and place the result in the variable I_LAST_NAME. Then the I_LAST_NAME variable can be used in another program. This program uses a special PL/SQL procedure named DBMS_OUTPUT.PUT_LINE to display the contents of the variable.

This procedure uses exactly one argument and produces a single line of text containing this argument as output. To see the contents of the line (the output), you must execute the following command:

```
SET SERVEROUTPUT ON
```

You can type this command at the SQL prompt in SQL*Plus. However, it is safer to include this command in the PL/SQL program to make sure that it will be executed when you run the program.

The complete program appears in Figure 8.2. It begins with the SET SERVEROUTPUT ON command. Next is the word DECLARE followed by the declaration of any variables. To declare a variable, you indicate the name of the variable and its type. In these programs, the variable names consist of the corresponding item in the database, preceded by the letter I and an underscore (I_). The name of the variable corresponding to the REP_NUM column, for example, is I_REP_NUM.

FIGURE 8.2
....
PL/SQL program
with a SELECT
command

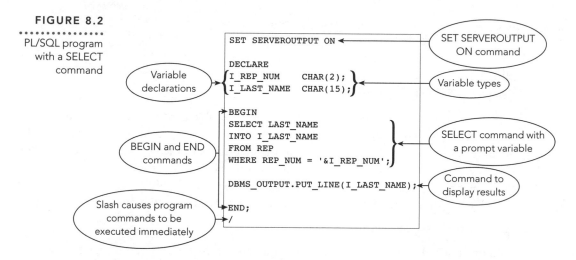

Note: Variable names in PL/SQL must start with a letter and can contain letters, dollar signs, underscores, and number signs. Variable names cannot exceed 30 characters.

Note: PL/SQL commands, like SQL commands, are free-format and can include blank lines to separate important sections of the program and spaces on the lines to make the commands more readable.

The **procedural code**, the commands that specify exactly what the program is to do, appears between the BEGIN and END commands. In this example, the procedural code begins with the SQL command to place the result in a variable. The next command uses the DBMS_OUTPUT.PUT_LINE procedure to display the contents of the I_LAST_NAME variable. Notice that each variable declaration and command and the word END are followed by semicolons.

The slash (/) at the end of the program appears on its own line. When you run the script file containing this program, the slash causes the commands in the program to be executed immediately. Without the slash, the commands would only be loaded into memory and you would need to type the slash at the SQL prompt to run the program. Most programs include the slash so the user doesn't need to type it.

Figure 8.3 shows the results of running this program and entering the value 20 for the rep number. Oracle displays the old line, the new line (with the new value inserted), and the result. In this case, the result contains the name Kaiser, which is the last name of sales rep 20.

FIGURE 8.3   Results of running the PL/SQL command

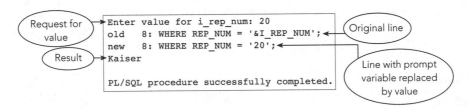

Using the %TYPE Attribute

When declaring variables, you can also ensure that a variable has the same type as a particular column in a table by using the %TYPE attribute. To do so, you include the name of the table, followed by a period and the name of the column, and then %TYPE. When you use %TYPE, you do not enter a data type because the variable is automatically assigned the same type as the corresponding column. To assign the variable I_REP_NUM the same type as the REP_NUM column in the REP table, for example, you would use the following declaration:

```
I_REP_NUM REP.REP_NUM%TYPE
```

Figure 8.4 is the same program as the one shown in Figure 8.2, except for the new variable declarations.

FIGURE 8.4

Using the %TYPE attribute

```
SET SERVEROUTPUT ON

DECLARE
I_REP_NUM      REP.REP_NUM%TYPE;
I_LAST_NAME    REP.LAST_NAME%TYPE;

BEGIN
SELECT LAST_NAME
INTO I_LAST_NAME
FROM REP
WHERE REP_NUM = '&I_REP_NUM';

DBMS_OUTPUT.PUT_LINE(I_LAST_NAME);

END;
/
```

Assigns I_REP_NUM the same type as the REP_NUM column in the REP table

Retrieving a Single Row from a Join

You can use an embedded SQL command to join tables, as illustrated in Example 3.

EXAMPLE 3 : Obtain the name of the customer whose customer number is stored in the prompt variable I_CUSTOMER_NUM, and the last and first names of the sales rep who represents this customer.

This query involves joining the CUSTOMER and REP tables. Because the restriction involves the primary key of the CUSTOMER table and each customer is related to exactly one sales rep, the query produces a single row. The method for handling this query is similar to that of the preceding queries. The corresponding program is shown in Figure 8.5. Notice that three DBMS_OUTPUT.PUT_LINE commands are required to display the customer's name and the rep's last and first names.

FIGURE 8.5
• • • • • • • • • • • • • • •
PL/SQL program
containing a join

```
SET SERVEROUTPUT ON

DECLARE
I_LAST_NAME        REP.LAST_NAME%TYPE;
I_FIRST_NAME       REP.FIRST_NAME%TYPE;
I_CUSTOMER_NUM     CUSTOMER.CUSTOMER_NUM%TYPE;
I_CUSTOMER_NAME    CUSTOMER.CUSTOMER_NAME%TYPE;

BEGIN
SELECT LAST_NAME, FIRST_NAME, CUSTOMER_NAME
INTO I_LAST_NAME, I_FIRST_NAME, I_CUSTOMER_NAME
FROM REP, CUSTOMER
WHERE REP.REP_NUM = CUSTOMER.REP_NUM
AND CUSTOMER_NUM = '&I_CUSTOMER_NUM';

DBMS_OUTPUT.PUT_LINE(I_CUSTOMER_NAME);
DBMS_OUTPUT.PUT_LINE(I_LAST_NAME);
DBMS_OUTPUT.PUT_LINE(I_FIRST_NAME);

END;
/
```

Joins the REP and CUSTOMER tables →

Displays the customer's name, and the rep's last and first names →

The results of running this program are shown in Figure 8.6.

FIGURE 8.6
• • • • • • • • • • • • • • •
Result of running
the PL/SQL
program

```
Enter value for i_customer_num: 148
old  12: AND CUSTOMER_NUM = '&I_CUSTOMER_NUM';
new  12: AND CUSTOMER_NUM = '148';
Al's Appliance and Sport
Kaiser
Valerie

PL/SQL procedure successfully completed.
```

Inserting a Row into a Table

When you are updating databases from within PL/SQL programs, you use appropriate SQL commands. For example, to add a row to a table in the database, you use the SQL INSERT command, as illustrated in Example 4.

EXAMPLE 4 : Add a row to the REP table. Use prompt variables to obtain values for the fields.

Figure 8.7 shows a program to add the row. The values in the INSERT command are prompt variables.

FIGURE 8.7 PL/SQL program to insert a row

```
DECLARE
I_REP_NUM      REP.REP_NUM%TYPE;
I_LAST_NAME    REP.LAST_NAME%TYPE;
I_FIRST_NAME   REP.FIRST_NAME%TYPE;
I_STREET       REP.STREET%TYPE;
I_CITY         REP.CITY%TYPE;
I_STATE        REP.STATE%TYPE;
I_ZIP          REP.ZIP%TYPE;
I_COMMISSION   REP.COMMISSION%TYPE;
I_RATE         REP.RATE%TYPE;

BEGIN
INSERT INTO REP (REP_NUM, LAST_NAME, FIRST_NAME, STREET, CITY, STATE, ZIP, COMMISSION, RATE)
VALUES
('&I_REP_NUM', '&I_LAST_NAME', '&I_FIRST_NAME', '&I_STREET', '&I_CITY', '&I_STATE', '&I_ZIP',
&I_COMMISSION, &I_RATE);

END;
/
```

Values come from prompt variables

INSERT command

Figure 8.8 shows the results of executing the program. First the program requests values for the rep number, last name, first name, street, city, state, and zip. These are the items on the first line of values in the INSERT command; your command's format might differ. It then displays the old and new versions of this line. Next, it asks for values for commission and rate, the items on the second line of values in the INSERT command in Figure 8.7. It displays old and new versions of this second line of values and then executes the completed INSERT command, adding a new record to the REP table.

FIGURE 8.8 Running the PL/SQL program

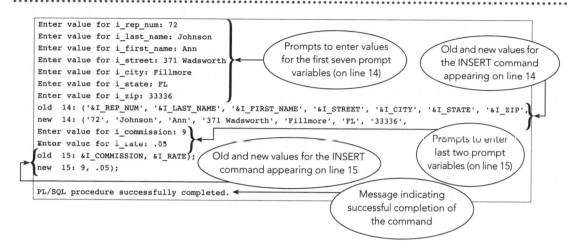

Changing a Single Row in a Table

Just as you used SQL commands to insert rows in a database, you can also use them to update the rows, as illustrated in Example 5.

EXAMPLE 5 :   Change the last name of the sales rep whose number is stored in I_REP_NUM to the value currently stored in I_LAST_NAME.

Again, the only difference between this example and the update examples in Chapter 5 is the use of variables. The program appears in Figure 8.9.

FIGURE 8.9

PL/SQL command to update rows

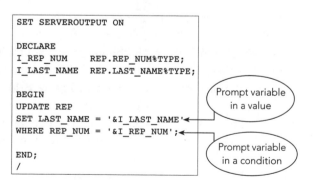

When you run this program, Oracle will first ask you for a last name. After you enter the last name, it will ask you for a rep number. Once you have entered a rep number, Oracle will change the last name of the rep whose number is equal to the number you entered to the name you entered.

Deleting Rows from a Table

Just as you would expect, if you must use SQL commands to insert and change rows in a table, you also must use SQL commands to delete rows, as illustrated in Example 6.

EXAMPLE 6 : Delete the sales rep whose number currently is stored in I_REP_NUM from the REP table.

The program appears in Figure 8.10. When you run this program, Oracle will ask you for a rep number. Once you enter a rep number, Oracle will delete the row for that rep.

FIGURE 8.10
. . . .
PL/SQL
command to
delete rows

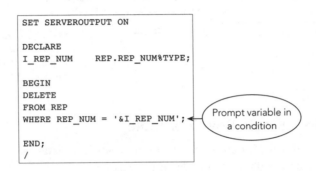

```
SET SERVEROUTPUT ON

DECLARE
I_REP_NUM      REP.REP_NUM%TYPE;

BEGIN
DELETE
FROM REP
WHERE REP_NUM = '&I_REP_NUM';

END;
/
```

Prompt variable in a condition

Deleting Rows from Multiple Tables

Sometimes you need to delete more than one row from a table. If you delete an order from the ORDERS table, for example, you also need to delete all associated order lines from the ORDER_LINE table, as illustrated in Example 7.

EXAMPLE 7 : Delete the order whose number is stored in I_ORDER_NUM from the ORDERS table, and then delete each order line for the order whose order number is currently stored in the variable from the ORDER_LINE table.

A program to accomplish this task appears in Figure 8.11. Because both DELETE commands use the *same* order number, you should not use prompt variables in the DELETE commands, because users would need to enter the order number twice. Not only is this less convenient for the user, it is possible for the user to enter *different* order numbers in error.

A better way is to have the user enter a single order number into a variable prior to the program's execution of the DELETE commands. Because you can think of this other variable as a work variable, it is common to begin such a variable name with the letter W (for example, W_ORDER_NUM). In the program shown in Figure 8.11, the first command requests the user to enter a value for W_ORDER_NUM, the prompt variable, and then places this value in the variable I_ORDER_NUM. The DELETE commands then use the variable I_ORDER_NUM to delete the rows in the ORDERS and ORDER_LINE tables. Notice that there are no ampersands (&) in the DELETE commands. In addition, single quotation marks are not required for I_ORDER_NUM in the DELETE command because it is not a prompt variable.

FIGURE 8.11
· · · · · · · · · · · · · · ·
PL/SQL
command to
delete rows from
multiple tables

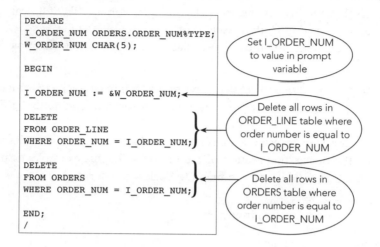

```
DECLARE
I_ORDER_NUM ORDERS.ORDER_NUM%TYPE;
W_ORDER_NUM CHAR(5);

BEGIN

I_ORDER_NUM := &W_ORDER_NUM;

DELETE
FROM ORDER_LINE
WHERE ORDER_NUM = I_ORDER_NUM;

DELETE
FROM ORDERS
WHERE ORDER_NUM = I_ORDER_NUM;

END;
/
```

Set I_ORDER_NUM to value in prompt variable

Delete all rows in ORDER_LINE table where order number is equal to I_ORDER_NUM

Delete all rows in ORDERS table where order number is equal to I_ORDER_NUM

When you run this program, Oracle will request a value for W_ORDER_NUM. The program then sets I_ORDER_NUM to the entered value. Then it deletes rows from the ORDERS and ORDER_LINE tables containing the order number.

■ Multiple-Row Select

The earlier SELECT example posed no problems for PL/SQL because the SELECT commands retrieved only individual rows. UPDATE commands that update multiple rows pose no special difficulty nor do DELETE commands that delete multiple rows. The commands are executed and the updates or deletions occur. Then the program can move on to the next task.

What happens when a SELECT command retrieves multiple rows? What if, for example, the SELECT command retrieves the number and name of each customer represented by the sales rep whose number is stored in I_REP_NUM? There is a problem—PL/SQL can process only one record at a time, but this SQL command produces multiple rows. Whose number and name is placed in I_CUSTOMER_NUM and I_CUSTOMER_NAME if 100 customers are retrieved? Should you make I_CUSTOMER_NUM and I_CUSTOMER_NAME arrays capable of holding 100 customers and, if so, what should be the size of these arrays? Fortunately, you can solve this problem by using a cursor.

Using Cursors

A **cursor** is a pointer to a row in the collection of rows retrieved by a SQL command. (This is *not* the same cursor that you see on your computer screen.) The cursor advances one row at a time to provide sequential, one-record-at-a-time access to the retrieved rows so PL/SQL can process the rows. By using a cursor, PL/SQL can process the set of retrieved rows as though they were records in a sequential file.

To use a cursor, you must first declare it, as illustrated in Example 8.

EXAMPLE 8 : Retrieve and list the number and name of each customer represented by the sales rep whose number is stored in the variable I_REP_NUM.

The first step in using a cursor is to declare the cursor and describe the associated query in the declaration section of the program. In this example, assuming the cursor is named CUSTGROUP, the command to declare the cursor is:

```
CURSOR CUSTGROUP IS
SELECT CUSTOMER_NUM, CUSTOMER_NAME
FROM CUSTOMER
WHERE REP_NUM = '&I_REP_NUM';
```

This command does *not* cause the query to be executed at this time; it only declares a cursor named CUSTGROUP and associates the cursor with the indicated query. Using a cursor in the procedural portion of the program involves three commands: OPEN, FETCH, and CLOSE. The **OPEN** command opens the cursor and causes the query to be executed, making the results available to the program. Executing a **FETCH** command advances the cursor to the next row in the set of rows retrieved by the query and places the contents of the row in the indicated variables. Finally, the **CLOSE** command closes a cursor and deactivates it. Data retrieved by the execution of the query is no longer available. The cursor could be opened again later and processing could begin again.

The OPEN, FETCH, and CLOSE commands used in processing a cursor are analogous to the OPEN, READ, and CLOSE commands used in processing a sequential file. Next, you will examine how each of these commands is coded in PL/SQL.

Opening a Cursor

Prior to opening the cursor, there are no rows available to be fetched. In Figure 8.12, this is indicated by the absence of data in the CUSTGROUP portion of the figure. The right side of the figure illustrates the variables into which the data will be placed (I_CUSTOMER_NUM and I_CUSTOMER_NAME) and the special value CUSTGROUP%NOTFOUND. Once the cursor has been opened and all the records have been fetched, the CUSTGROUP%NOTFOUND value is set to TRUE. Programs using the cursor can use this value to indicate when the fetching of rows is complete.

FIGURE 8.12 Before OPEN

CUSTGROUP

CUSTOMER_ NUM	CUSTOMER_NAME		I_CUSTOMER_ NUM	I_CUSTOMER_NAME	CUSTGROUP %NOTFOUND
		←—no row to be fetched			FALSE

The OPEN command is written as follows:

```
OPEN CUSTGROUP;
```

Figure 8.13 shows the result of opening the CUSTGROUP cursor. In the figure, assume that I_REP_NUM is set to 20 before the OPEN command is executed; there are now three rows available to be fetched. No rows have yet been fetched, as indicated by the absence of values in I_CUSTOMER_NUM and I_CUSTOMER_NAME. CUSTGROUP%NOTFOUND is still FALSE. The cursor is positioned at the first row; that is, the next FETCH command causes the contents of the first row to be placed in the indicated variables.

FIGURE 8.13 After OPEN, but before first FETCH

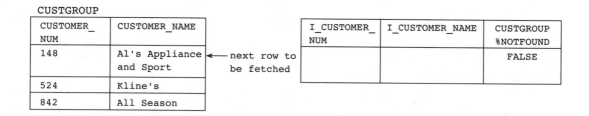

CUSTGROUP

CUSTOMER_ NUM	CUSTOMER_NAME		I_CUSTOMER_ NUM	I_CUSTOMER_NAME	CUSTGROUP %NOTFOUND
148	Al's Appliance and Sport	←—next row to be fetched			FALSE
524	Kline's				
842	All Season				

Fetching Rows from a Cursor

To fetch (get) the next row from a cursor, use the FETCH command. The FETCH command is written as follows:

```
FETCH CUSTGROUP INTO I_CUSTOMER_NUM, I_CUSTOMER_NAME;
```

Note that the INTO clause is associated with the FETCH command itself and not with the query used in the cursor definition. The execution of this query probably produces multiple rows. The execution of the FETCH command produces only a single row, so it is appropriate that the FETCH command causes data to be placed in the indicated variables.

Figures 8.14 through 8.17 show the result of four FETCH commands. The first three fetches are successful. In each case, the data from the appropriate row in the cursor is placed in the indicated variables and CUSTGROUP%NOTFOUND is still FALSE. The fourth FETCH command is different, however, because there is no more data to fetch. In this case, the contents of the variables are left untouched and CUSTGROUP%NOTFOUND is set to TRUE.

FIGURE 8.14 After first FETCH

CUSTGROUP

CUSTOMER_NUM	CUSTOMER_NAME
148	Al's Appliance and Sport
524	Kline's
842	All Season

←next row to be fetched (pointing at 524)

I_CUSTOMER_NUM	I_CUSTOMER_NAME	CUSTGROUP%NOTFOUND
148	Al's Appliance and Sport	FALSE

FIGURE 8.15 After second FETCH

CUSTGROUP

CUSTOMER_NUM	CUSTOMER_NAME
148	Al's Appliance and Sport
524	Kline's
842	All Season

←next row to be fetched (pointing at 842)

I_CUSTOMER_NUM	I_CUSTOMER_NAME	CUSTGROUP%NOTFOUND
524	Kline's	FALSE

FIGURE 8.16 After third FETCH

CUSTGROUP

CUSTOMER_NUM	CUSTOMER_NAME
148	Al's Appliance and Sport
524	Kline's
842	All Season

←next row to be fetched (pointing at empty row)

I_CUSTOMER_NUM	I_CUSTOMER_NAME	CUSTGROUP%NOTFOUND
842	All Season	FALSE

FIGURE 8.17   After attempting a fourth FETCH (CUSTGROUP%NOTFOUND is TRUE)

CUSTGROUP

CUSTOMER_NUM	CUSTOMER_NAME
148	Al's Appliance and Sport
524	Kline's
842	All Season

← no more rows to be fetched

I_CUSTOMER_NUM	I_CUSTOMER_NAME	CUSTGROUP %NOTFOUND
842	All Season	TRUE

Closing a Cursor

The CLOSE command is written as follows:

```
CLOSE CUSTGROUP;
```

Figure 8.18 shows the result of closing the CUSTGROUP cursor. The data is no longer available.

FIGURE 8.18

After CLOSE

CUSTGROUP

CUSTOMER_NUM	CUSTOMER_NAME

← no rows to be fetched

Complete Program Using a Cursor

Figure 8.19 shows a complete program using a cursor. The declaration portion contains the cursor definition. The procedural portion begins with the command to open the CUSTGROUP cursor. The statements between the LOOP and END LOOP commands create a loop that begins by fetching the next row from the cursor and placing the results in I_CUSTOMER_NUM and I_CUSTOMER_NAME. The EXIT command tests to see whether the condition CUSTGROUP%NOTFOUND is TRUE; if it is, the loop is terminated. If it is not, the DBMS_OUTPUT.PUT_LINE commands display the contents of I_CUSTOMER_NUM and I_CUSTOMER_NAME.

FIGURE 8.19
....
PL/SQL program
with a cursor

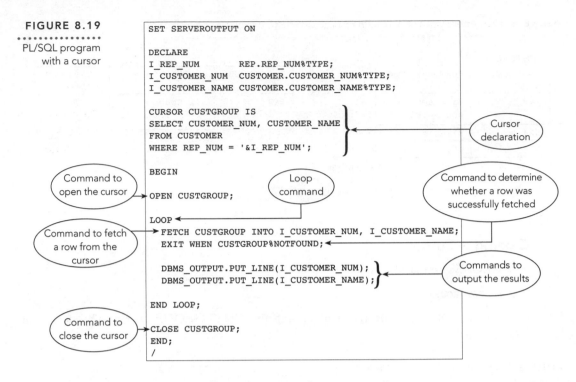

```
SET SERVEROUTPUT ON

DECLARE
I_REP_NUM         REP.REP_NUM%TYPE;
I_CUSTOMER_NUM    CUSTOMER.CUSTOMER_NUM%TYPE;
I_CUSTOMER_NAME   CUSTOMER.CUSTOMER_NAME%TYPE;

CURSOR CUSTGROUP IS
SELECT CUSTOMER_NUM, CUSTOMER_NAME
FROM CUSTOMER
WHERE REP_NUM = '&I_REP_NUM';

BEGIN

OPEN CUSTGROUP;

LOOP
    FETCH CUSTGROUP INTO I_CUSTOMER_NUM, I_CUSTOMER_NAME;
    EXIT WHEN CUSTGROUP%NOTFOUND;

    DBMS_OUTPUT.PUT_LINE(I_CUSTOMER_NUM);
    DBMS_OUTPUT.PUT_LINE(I_CUSTOMER_NAME);

END LOOP;

CLOSE CUSTGROUP;
END;
/
```

Cursor declaration

Command to open the cursor

Loop command

Command to determine whether a row was successfully fetched

Command to fetch a row from the cursor

Commands to output the results

Command to close the cursor

Figure 8.20 shows the results of running the program shown in Figure 8.19. After the user enters 20 as the value for the rep number, Oracle displays the old and new versions of the corresponding line in the program, and then displays the number and name of each customer of sales rep 20.

FIGURE 8.20
....
Results of
running PL/SQL
program

```
Enter value for i_rep_num: 20
old    9: WHERE REP_NUM = '&I_REP_NUM';
new    9: WHERE REP_NUM = '20';
148
Al's Appliance and Sport
524
Kline's
842
All Season

PL/SQL procedure successfully completed.
```

More Complex Cursors

The query formulation that defined the cursor in Example 8 was simple. Any SQL query is legitimate in a cursor definition. In fact, the more complicated the requirements for retrieval, the more numerous the benefits derived by the programmer who uses embedded SQL. Consider the query in Example 9.

EXAMPLE 9 For each order that contains an order line for the part whose part number is stored in I_PART_NUM, retrieve the order number, order date, customer number, name of the customer that placed the order, and the last and first names of the sales rep who represents the customer.

Opening and closing the cursor is done exactly as shown in Example 8. The only difference in the FETCH command is that a different set of variables is used in the INTO clause. Thus, the only real difference is the cursor definition. The program shown in Figure 8.21 contains the appropriate cursor definition.

FIGURE 8.21 PL/SQL program with cursor that involves joining multiple tables

```
SET SERVEROUTPUT ON

DECLARE
I_ORDER_NUM ORDERS.ORDER_NUM%TYPE;
I_ORDER_DATE ORDERS.ORDER_DATE%TYPE;
I_CUSTOMER_NUM ORDERS.CUSTOMER_NUM%TYPE;
I_CUSTOMER_NAME CUSTOMER.CUSTOMER_NAME%TYPE;
I_LAST_NAME REP.LAST_NAME%TYPE;
I_FIRST_NAME REP.FIRST_NAME%TYPE;

CURSOR ORDGROUP IS
SELECT ORDERS.ORDER_NUM, ORDER_DATE, ORDERS.CUSTOMER_NUM, CUSTOMER_NAME,
LAST_NAME, FIRST_NAME
FROM ORDER_LINE, ORDERS, CUSTOMER, REP
WHERE ORDER_LINE.ORDER_NUM = ORDERS.ORDER_NUM
AND ORDERS.CUSTOMER_NUM = CUSTOMER.CUSTOMER_NUM
AND CUSTOMER.REP_NUM = REP.REP_NUM
AND PART_NUM = '&I_PART_NUM';

BEGIN

OPEN ORDGROUP;
LOOP
  FETCH ORDGROUP INTO I_ORDER_NUM, I_ORDER_DATE, I_CUSTOMER_NUM,
  I_CUSTOMER_NAME, I_LAST_NAME, I_FIRST_NAME;
  EXIT WHEN ORDGROUP%NOTFOUND;

  DBMS_OUTPUT.PUT_LINE(I_ORDER_NUM);
  DBMS_OUTPUT.PUT_LINE(I_ORDER_DATE);
  DBMS_OUTPUT.PUT_LINE(I_CUSTOMER_NUM);
  DBMS_OUTPUT.PUT_LINE(I_CUSTOMER_NAME);
  DBMS_OUTPUT.PUT_LINE(I_LAST_NAME);
  DBMS_OUTPUT.PUT_LINE(I_FIRST_NAME);

END LOOP;

CLOSE ORDGROUP;
END;
/
```

The results of running this program are shown in Figure 8.22. After the user enters DR93 as the value for the part number, Oracle displays the old and new versions of the corresponding line in the program, and then displays the requested data for each order containing an order line for part DR93.

FIGURE 8.22
· · · · · · · · · · · · · · · ·
Results of
running PL/SQL
program

```
Enter value for i_part_num: DR93
old   16: AND PART_NUM = '&I_PART_NUM';
new   16: AND PART_NUM = 'DR93';
21610
20-OCT-07
356
Ferguson's
Perez
Juan
21619
23-OCT-07
148
Al's Appliance and Sport
Kaiser
Valerie

PL/SQL procedure successfully completed.
```

Advantages of Cursors

The retrieval requirements in Example 9 are complicated. Beyond coding the preceding cursor definition, the programmer doesn't need to worry about the mechanics of obtaining the necessary data or placing it in the right order, because this happens automatically when the cursor is opened. To the programmer, it seems as if a sequential file already existed with precisely the right data in it, sorted in the right order. This assumption leads to three main advantages:

- The coding in the program is greatly simplified.

- In a normal PL/SQL program, the programmer must determine the most efficient way to access the data. In a program using embedded SQL, a special component of the database management system called the **optimizer** determines the best way to access the data. The programmer isn't concerned with the best way to retrieve the data. In addition, if an underlying structure changes (for example, an additional index is created), the optimizer determines the best way to execute the query in light of the new structure. The program does not have to change at all.

- If the database structure changes in such a way that the necessary information is still obtainable using a different query, the only change required in the program is the cursor definition. The procedural code is not affected.

Updating Cursors

You can update the rows encountered in processing cursors by including an additional clause—FOR UPDATE OF—in the cursor definition. For example, consider the update requirement in Example 10.

EXAMPLE 10 Increase the credit limit by $200 for each customer represented by the sales rep whose number is currently stored in the variable I_REP_NUM, whose balance is not over the credit limit, and whose credit limit is $10,000 or more. Increase the credit limit by $100 for each customer of this sales rep whose balance is not over the credit limit and whose credit limit is less than $10,000. List the number and name of each customer of this sales rep whose balance is greater than the credit limit.

The program is shown in Figure 8.23. To update the credit limits, you must include the FOR UPDATE OF CREDIT_LIMIT clause in the cursor definition. The procedural code for opening, closing, and fetching rows from the cursor is the same one as in the previous examples.

FIGURE 8.23   PL/SQL program that updates data through a cursor

```
SET SERVEROUTPUT ON

DECLARE

I_REP_NUM          CUSTOMER.REP_NUM%TYPE;
I_CUSTOMER_NUM     CUSTOMER.CUSTOMER_NUM%TYPE;
I_CUSTOMER_NAME    CUSTOMER.CUSTOMER_NAME%TYPE;
I_BALANCE          CUSTOMER.BALANCE%TYPE;
I_CREDIT_LIMIT     CUSTOMER.CREDIT_LIMIT%TYPE;

CURSOR CREDGROUP IS
SELECT CUSTOMER_NUM, CUSTOMER_NAME, BALANCE, CREDIT_LIMIT
FROM CUSTOMER
WHERE REP_NUM = '&I_REP_NUM'
FOR UPDATE OF CREDIT_LIMIT;

BEGIN

OPEN CREDGROUP;
LOOP
  FETCH CREDGROUP INTO I_CUSTOMER_NUM, I_CUSTOMER_NAME, I_BALANCE, I_CREDIT_LIMIT;
  EXIT WHEN CREDGROUP%NOTFOUND;

  IF I_BALANCE > I_CREDIT_LIMIT THEN
      DBMS_OUTPUT.PUT_LINE(I_CUSTOMER_NUM);
      DBMS_OUTPUT.PUT_LINE(I_CUSTOMER_NAME);
    ELSE
      IF I_CREDIT_LIMIT >= 10000 THEN
          UPDATE CUSTOMER
            SET CREDIT_LIMIT = CREDIT_LIMIT + 200
            WHERE CURRENT OF CREDGROUP;
        ELSE
          UPDATE CUSTOMER
            SET CREDIT_LIMIT = CREDIT_LIMIT + 100
            WHERE CURRENT OF CREDGROUP;
      END IF;
  END IF;

END LOOP;

CLOSE CREDGROUP;
END;
/
```

Cursor declaration contains FOR UPDATE OF clause

Tests to determine whether the balance is greater than the credit limit

If the balance is greater than the credit limit, output the number and name

UPDATE commands

Tests to determine whether the credit limit is greater than or equal to $10,000

Commands contain WHERE CURRENT OF clause

The procedural code includes an IF statement to update the row that was fetched. First the credit limit is compared with the balance. If the balance is larger, the customer number and name are displayed. If the balance is not larger, then the credit limit is compared with $10,000. If the credit limit is equal to or greater than $10,000, then the row is updated by adding $200 to the credit limit. If the credit limit is less than $10,000, then the row is updated by adding $100 to the credit limit. The WHERE CURRENT OF CREDGROUP clause indicates that the update is to apply only to the row just fetched. Without this clause, and in the absence of a WHERE clause to restrict the scope of the update, this command would update *every* customer's credit limit.

■ Error Handling

Programs must be able to handle exceptional conditions that can arise when accessing the database. For example, in the program shown in Figure 8.2, the user entered a rep number and the program displayed the corresponding last name. What if the user entered a rep number that did not exist? This situation would cause an error because Oracle would not find any last name to display.

You can handle errors by using the EXCEPTION command as illustrated in Figure 8.24. This program attempts to find the name of the sales rep who has a particular commission rate. It is possible to enter a rate that does not correspond to any sales rep (for example, no rep has the rate 0.06), resulting in a NO_DATA_FOUND error. Likewise, it is possible to enter a rate that corresponds to more than one sales rep (for example, two reps have the rate 0.05), resulting in a TOO_MANY_ROWS error. Following the word EXCEPTION is a command indicating that if the NO_DATA_FOUND error occurs, the program is to display the message "No records selected." If the TOO_MANY_ROWS error occurs, the program is to display the message "Too many records." In either case, the program cannot display the last name of the rep. If you run the program and enter 0.06 as the rate, you will see the "No records selected" message. If you enter 0.05 as the rate, you will see the "Too many records" message.

FIGURE 8.24
....
PL/SQL program
with error
handling

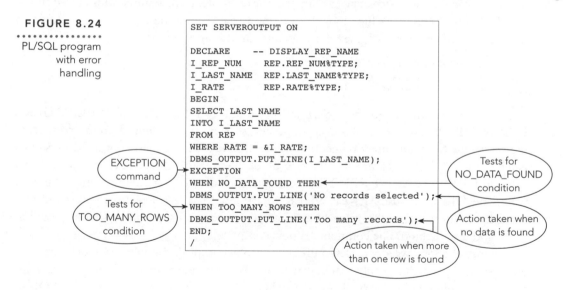

■ Using SQL in Microsoft Access Programs

Not every programming language accepts SQL commands as readily as PL/SQL. For example, in Microsoft Access, programs are written in Visual Basic, which does not allow the inclusion of SQL commands directly in the code. If the SQL command is stored in a string variable, however, you can use the DoCmd.RunSQL command to run the command. The procedure in which you place the SQL command can include **arguments**, which are values that provide information to the procedure. For example, in a procedure to delete a sales rep, the argument would be the sales rep number.

Deleting Rows

To delete the sales rep whose number is 20, the command is:

```
DELETE FROM REP WHERE REP_NUM = '20';
```

When you write this type of command, you usually don't know in advance the specific sales rep number that you want to delete; it would be passed as an argument to the procedure containing this DELETE command. In the following example, the sales rep number is stored in an argument named I_REP_NUM.

EXAMPLE 11 : Delete from the REP table the sales rep whose number currently is stored in I_REP_NUM.

Usually statements in the procedure create the appropriate DELETE command, using the value in any necessary arguments. For example, if the command is stored in the variable named strSQL (which must be a string variable) and the rep number is stored in the argument I_REP_NUM, the following command is appropriate:

```
strSQL = "DELETE FROM REP WHERE REP_NUM = '"
strSQL = strSQL & I_REP_NUM
strSQL = strSQL & "';"
```

The first command sets the strSQL string variable to DELETE FROM REP WHERE REP_NUM = '; that is, it creates everything necessary in the command up to and including the single quotation mark preceding the rep number. The second command uses concatenation (&). It changes strSQL to the result of the previous value concatenated with the value in I_REP_NUM. If I_REP_NUM contains the value 20, for example, the command would be DELETE FROM REP WHERE REP_NUM = '20. The final command sets strSQL to the result of the value already created, concatenated with a single quotation mark and a semicolon. The command is now complete.

Figure 8.25 shows a completed procedure to accomplish the necessary deletion. You enter such a procedure in the Microsoft Visual Basic window. In the program, the Dim statement creates a string variable named strSQL. The next three commands set strSQL to the appropriate SQL command. Finally, the DoCmd.RunSQL command runs the SQL command stored in strSQL.

FIGURE 8.25   Code to delete a sales rep

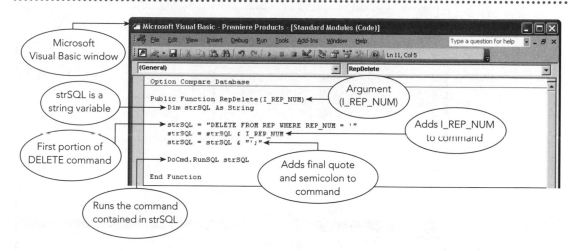

Note: If you have any doubts about the way you have constructed the SQL command in strSQL, you can include the Debug.Print (strSQL) command after the set of commands that construct strSQL. This command displays the entire command before it is executed so you can review it for accuracy. If you need to correct an error, rerun the program after making the necessary changes. If you get a strange error in your program, check your SQL command carefully to make sure that you concatenated it correctly.

Running the Code

Normally, you run code like the function shown in Figure 8.25 by calling it from another procedure or associating it with some event, such as clicking a button on a form. However, you can run it directly by using the Immediate window (click View on the menu bar, and then click Immediate Window to open it). Normally, you would use this window only for testing purposes, but you can use it here to see the result of running the code. To run a Function procedure, such as the one shown in Figure 8.25, in the Immediate window, type a question mark, followed by the name of the procedure and a set of parentheses, as shown in Figure 8.26. Place the values for any arguments in the parentheses. Assuming that you wanted to delete a sales rep whose number is 50, you would include "50" inside the parentheses as shown in the figure.

FIGURE 8.26
· · · · · · · · · · · · · · ·
Running the
code in the
Immediate
window

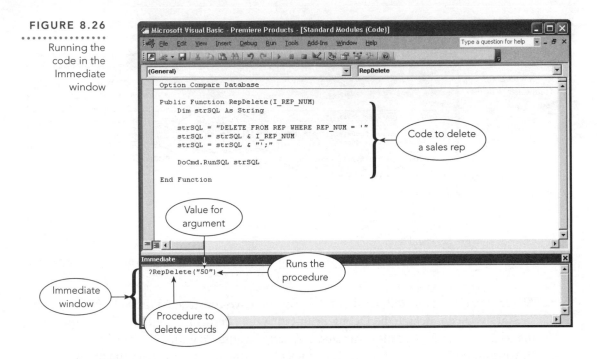

After you type the command and press the Enter key, the code will run and the appropriate action will occur. In this case, the command deletes the sales rep with the number 50 (assuming there is a sales rep 50).

Updating Rows

A procedure that updates a table using an UPDATE command is similar to the one used to delete a sales rep. In Example 12, two arguments are required. One of them, I_LAST_NAME, contains the new name for the sales rep. The other, I_REP_NUM, contains the number of the rep whose name is to be changed.

EXAMPLE 12 : Change the last name of the sales rep whose number is stored in
    : I_REP_NUM to the value currently stored in I_LAST_NAME.

This example is similar to the previous one with two important differences. First, you need the UPDATE command rather than the DELETE command. Second, there are two arguments rather than one, so there are two portions of the construction of the SQL command that involve variables. The complete procedure is shown in Figure 8.27.

FIGURE 8.27   Code to change a rep's last name

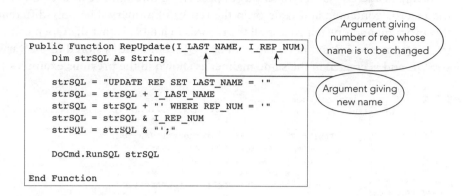

```
Public Function RepUpdate(I_LAST_NAME, I_REP_NUM)
    Dim strSQL As String

    strSQL = "UPDATE REP SET LAST_NAME = '"
    strSQL = strSQL + I_LAST_NAME
    strSQL = strSQL + "' WHERE REP_NUM = '"
    strSQL = strSQL & I_REP_NUM
    strSQL = strSQL & "';"

    DoCmd.RunSQL strSQL

End Function
```

Argument giving number of rep whose name is to be changed

Argument giving new name

To run this procedure, you would enter values for both arguments as shown in Figure 8.28.

FIGURE 8.28

Running the code to change a rep's last name

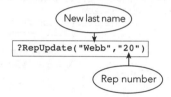

New last name

```
?RepUpdate("Webb","20")
```

Rep number

Inserting Rows

The process for inserting rows is similar. You create the appropriate INSERT command in the strSQL variable. There will be multiple arguments in the procedure—one for each value to be inserted.

Finding Rows

Just as when embedding SQL in PL/SQL, deleting or updating multiple rows causes no problems, because these procedures still represent a single operation, with all the work happening behind the scenes. A SELECT command that returns several rows, however, poses serious problems for record-at-a-time languages like PL/SQL and Visual Basic. You handle SELECT commands differently than in PL/SQL. In particular, there are no cursors. Instead, you handle the results of a query just as you would use a loop to process through the records in a table.

EXAMPLE 13 : Retrieve and list the number and name of each customer represented by the sales rep whose number is stored in the variable I_REP_NUM.

Figure 8.29 shows a procedure to accomplish the indicated task. The statements involving rs and cnn are a typical way of processing through a recordset, that is, through all the records contained in a table or in the results of a query. The only difference between this program and one to process all the records in a table is that the Open command refers to a SQL command and not a table. (The SQL command is stored in the variable named strSQL and is created in the same manner as shown in the previous examples.)

FIGURE 8.29 Code to find customers of a specific rep

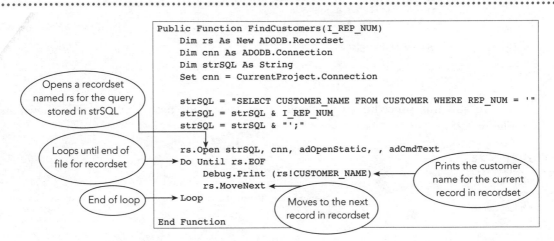

The loop continues until reaching the end of file for the recordset, that is, until all records have been processed. Within the loop, you can use the Debug.Print command to print a value. In this case, the value to be printed is rs!CUSTOMER_NAME. This indicates the contents of the CUSTOMER_NAME column for the record in the recordset (rs) on which Access is currently positioned. The next command, rs.MoveNext, moves to the next record in the recordset. The loop continues until all records in the recordset have been processed.

Figure 8.30 shows the results of running this procedure and entering a value of "35" as an argument. Access displays the four customers of sales rep 35.

FIGURE 8.30

Running the code to find customers of a sales rep

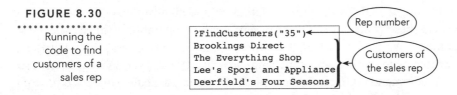

Note: Even if you expect only one record to be returned by the SQL query, you use the same process but would not need a loop.

In this chapter, you learned how to embed SQL commands in a procedural language. You created programs in PL/SQL with embedded SQL commands that retrieved single rows, inserted new rows, and changed and deleted existing rows. Then you examined the difficulties associated with embedded SQL commands that retrieve multiple rows and learned how to use cursors to update a database. You learned how to handle errors in PL/SQL programs that contain embedded SQL commands. Finally, you learned how to use SQL commands within Access even though it is not possible to embed the commands directly.

■ SUMMARY

- To embed a SQL command in a PL/SQL program, type the command in the procedural code.

- Variables in PL/SQL programs are declared after the word DECLARE. To assign variables the same type as a column in the database, use the %TYPE attribute.

- You can request user input in SQL commands by using prompt variables. To indicate that a variable is a prompt variable, precede the name with an ampersand (&).

- You can use SELECT commands as embedded SQL commands in PL/SQL programs only when a single row is retrieved.

- To place the results of a SELECT command in variables, use the INTO clause in the SELECT command.

- You can use INSERT, UPDATE, and DELETE commands in PL/SQL programs, even when they affect more than one row.

- If a SELECT command is to retrieve more than one row, it must be used to define a cursor that will be used to select one row at a time.

- To activate a cursor, use the OPEN command to execute the query in the cursor definition.

- To select the next row in PL/SQL, use the FETCH command.

- To deactivate a cursor, use the CLOSE command. The rows initially retrieved will no longer be available to PL/SQL.

- Data in the tables on which a cursor is based can be updated if the cursor definition includes the FOR UPDATE OF clause. To update data, include the WHERE CURRENT OF *cursor name* clause in the UPDATE command. This clause updates only the current (most recently fetched) row.

- To see whether an error has occurred, use the EXCEPTION command.

- To use SQL commands in Access, create the command in a string variable. To run the command stored in the string variable, use the DoCmd.RunSQL command.

- To process a collection of rows retrieved by a SELECT command in Access, use a recordset. Create the SQL command in a string variable and use the string variable in the command to open the recordset.

- To move to the next record in a recordset, use the MoveNext command.

■ KEY TERMS

argument
CLOSE
cursor
embed
FETCH

OPEN
optimizer
procedural code
prompt variable

■ REVIEW QUESTIONS

1. In which portion of a PL/SQL program do you embed SQL commands?

2. Where do you declare variables in PL/SQL programs?

3. How do you assign variables the same type as a column in the database?

4. How do you request input in SQL commands?

5. How do you indicate that a variable is a prompt variable?

6. Are there any conditions on SELECT commands that can be embedded in the procedural code in PL/SQL programs?

7. How do you place the results of a SELECT command into variables?

8. Can you use INSERT, UPDATE, or DELETE commands that affect more than one row in PL/SQL programs?

9. How do you use a SELECT command that retrieves more than one row in a PL/SQL program?

10. Which PL/SQL command activates a cursor?

11. Which PL/SQL command selects the next row in a cursor?

12. Which PL/SQL command deactivates a cursor?

13. Which clause can you use in a cursor definition to update data? Which clause must you use in the UPDATE command?

14. Which PL/SQL command examines a program to determine whether an error has occurred?

15. How do you use SQL commands in Access?

16. How do you process a collection of rows retrieved by a SELECT command in Access?

17. How do you move to the next record in a recordset?

■EXERCISES (Premiere Products)

1. Write PL/SQL programs to accomplish the following tasks:
   a. Obtain the name and credit limit of the customer whose number currently is stored in I_CUSTOMER_NUM. Place these values in the variables I_CUSTOMER_NAME and I_CREDIT_LIMIT, respectively. Print the contents of I_CUSTOMER_NAME and I_CREDIT_LIMIT.
   b. Obtain the order date, customer number, and name for the order whose number currently is stored in I_ORDER_NUM. Place these values in the variables I_ORDER_DATE, I_CUSTOMER_NUM and I_CUSTOMER_NAME, respectively. Print the contents of I_ORDER_DATE, I_CUSTOMER_NUM, and I_CUSTOMER_NAME.
   c. Add a row to the ORDERS table.
   d. Change the date of the order whose number is stored in I_ORDER_NUM to the date currently found in I_ORDER_DATE.
   e. Delete the order whose number is stored in I_ORDER_NUM.

2. Write PL/SQL programs to accomplish the following tasks:
   a. Retrieve and print the part number, part description, warehouse number, and unit price of every part in the item class stored in I_CLASS.
   b. Modify the program you created in Exercise 2a so that the program also updates the unit price. In particular, the program should update every part that is in warehouse 1 by adding 5% to the unit price and every part in warehouse 2 by adding 10% to the unit price.

3. Write Access functions to accomplish the following tasks:
   a. Delete the order whose number is stored in I_ORDER_NUM.
   b. Change the date of the order whose number is stored in I_ORDER_NUM to the date currently found in I_ORDER_DATE.
   c. Retrieve and print the part number, part description, warehouse number, and unit price of every part in the item class stored in I_CLASS.

1. Write PL/SQL programs to accomplish the following tasks:
   a. Obtain the first name and last name of the author whose number currently is stored in I_AUTHOR_NUM. Place these values in the variables I_AUTHOR_FIRST and I_AUTHOR_LAST. Print the contents of I_AUTHOR_NUM, I_AUTHOR_FIRST, and I_AUTHOR_LAST.
   b. Obtain the book title, publisher code, and publisher name for every book whose code currently is stored in I_BOOK_CODE. Place these values in the variables I_TITLE, I_PUBLISHER_CODE, and I_PUBLISHER_NAME, respectively. Print the contents of I_TITLE, I_PUBLISHER_CODE, and I_PUBLISHER_NAME.
   c. Add a row to the AUTHOR table.
   d. Change the last name of the author whose number is stored in I_AUTHOR_NUM to the value currently found in I_LAST_NAME.
   e. Delete the author whose number is stored in I_AUTHOR_NUM.

2. Write PL/SQL programs to accomplish the following tasks:
   a. Retrieve and print the book code, title, book type, and price for every book whose publisher code is stored in I_PUBLISHER_CODE.
   b. Modify the program you created in Exercise 2a so that the program also updates the book price. In particular, the program should increase the price by 4% for every book whose type is ART and by 3% for every book whose type is SFI.

3. Write Access functions to accomplish the following tasks:
   a. Delete the author whose number is stored in I_AUTHOR_NUM.
   b. Change the last name of the author whose number is stored in I_AUTHOR_NUM to the value currently found in I_AUTHOR_LAST.
   c. Retrieve and print the book code, title, book type, and price for every book whose publisher code is stored in I_PUBLISHER_CODE.

■EXERCISES (Alexamara Marina Group)

1. Write PL/SQL programs to accomplish the following tasks:

   a. Obtain the first name and last name of the owner whose number currently is stored in I_OWNER_NUM. Place these values in the variables I_FIRST_NAME and I_LAST_NAME. Print the contents of I_OWNER_NUM, I_FIRST_NAME, and I_LAST_NAME.

   b. Obtain the marina number, slip number, boat name, owner number, owner first name, and owner last name for the slip whose slip ID is currently stored in I_SLIP_ID. Place these values in the variables I_MARINA_NUM, I_SLIP_NUM, I_BOAT_NAME, I_OWNER_NUM, I_FIRST_NAME, and I_LAST_NAME, respectively. Print the contents of I_SLIP_ID, I_MARINA_NUM, I_SLIP_NUM, I_BOAT_NAME, I_OWNER_NUM, I_FIRST_NAME, and I_LAST_NAME.

   c. Add a row to the OWNER table.

   d. Change the last name of the owner whose number is stored in I_OWNER_NUM to the value currently found in I_LAST_NAME.

   e. Delete the owner whose number is stored in I_OWNER_NUM.

2. Write PL/SQL programs to accomplish the following tasks:

   a. Retrieve and print the marina number, slip number, rental fee, boat name, and owner number for every slip whose length is equal to the length stored in I_LENGTH.

   b. Modify the program you created in Exercise 2a so that the program also updates the rental fee. In particular, the program should increase the fee by 4% for every slip in marina 1 and by 3% for every slip in marina 2.

3. Write Access functions to accomplish the following tasks:

   a. Delete the owner whose number is stored in I_OWNER_NUM.

   b. Change the last name of the owner whose number is stored in I_OWNER_NUM to the value currently found in I_LAST_NAME.

   c. Retrieve and print the marina number, slip number, rental fee, boat name, and owner number for every slip whose length is equal to the length stored in I_LENGTH.

APPENDIX

SQL Reference

You can use this appendix to obtain details concerning important components and syntax of the SQL language. Items are arranged alphabetically. Each item contains a description, a reference to where the item is covered in the text, and, where appropriate, both an example and a description of the query results. Some SQL commands also include a description of the clauses associated with them. For each clause, there is a brief description and an indication of whether the clause is required or optional.

■ Aliases (Pages 114–119)

You can specify an alias (alternative name) for each table in a query. You can use the alias in the rest of the command by following the name of the table with a space and the alias name.

The following command creates an alias named R for the REP table and an alias named C for the CUSTOMER table:

```
SELECT R.REP_NUM, R.LAST_NAME, R.FIRST_NAME, C.CUSTOMER_NUM, C.CUSTOMER_NAME
FROM REP R, CUSTOMER C
WHERE R.REP_NUM = C.REP_NUM;
```

■ ALTER TABLE (Pages 148–152, 181–184)

Use the ALTER TABLE command to change a table's structure. As shown in Figure A-1, you type the ALTER TABLE command, followed by the table name, and then the alteration to perform.

FIGURE A-1   ALTER TABLE command

Clause	Description	Required?
ALTER TABLE *table name*	Indicates name of table to be altered.	Yes
alteration	Indicates type of alteration to be performed.	Yes

The following command alters the CUSTOMER table by adding a new CUSTOMER_TYPE column:

```
ALTER TABLE CUSTOMER
ADD CUSTOMER_TYPE CHAR(1);
```

The following command changes the CITY column in the CUSTOMER table so that it cannot accept nulls:

```
ALTER TABLE CUSTOMER
MODIFY CITY NOT NULL;
```

■ Column or Expression List (SELECT Clause) (Pages 75–76)

To select columns, use the SELECT clause followed by the list of columns, separated by commas.

The following SELECT clause selects the CUSTOMER_NUM, CUSTOMER_NAME, and BALANCE columns:

```
SELECT CUSTOMER_NUM, CUSTOMER_NAME, BALANCE
```

Use an asterisk in a SELECT clause to select all columns in a table. The following SELECT clause selects all columns:

```
SELECT *
```

Computed Columns (Pages 81–83)

You can use a computation in place of a column by typing the computation. For readability, you can type the computation in parentheses, although it is not necessary to do so.

The following SELECT clause selects the CUSTOMER_NUM and CUSTOMER_NAME columns as well as the results of subtracting the BALANCE column from the CREDIT_LIMIT column:

```
SELECT CUSTOMER_NUM, CUSTOMER_NAME, (CREDIT_LIMIT - BALANCE)
```

The DISTINCT Operator (Pages 89–91)

To avoid selecting duplicate values in a command, use the DISTINCT operator. If you omit the DISTINCT operator from the command and the same value appears on multiple rows in the table, that value will appear on multiple rows in the query results.

The following query selects all customer numbers from the ORDERS table, but it lists each customer number only once in the results:

```
SELECT DISTINCT(CUSTOMER_NUM)
FROM ORDERS;
```

Functions (Pages 87–91)

You can use functions in a SELECT clause. The most commonly used functions are AVG (to calculate an average), COUNT (to count the number of rows), MAX (to determine the maximum value), MIN (to determine the minimum value), and SUM (to calculate a total).

The following SELECT clause calculates the average balance:

```
SELECT AVG(BALANCE)
```

■ COMMIT (Pages 144–145)

Use the COMMIT command to make permanent any updates made since the last command. If no previous COMMIT command has been executed, the COMMIT command will make all the updates during the current work session permanent immediately. All updates become permanent automatically when you exit SQL. Figure A-2 describes the COMMIT command.

Clause	Description	Required?
COMMIT	Indicates that a COMMIT is to be performed.	Yes

The following command makes all updates since the most recent COMMIT command permanent:

```
COMMIT;
```

Conditions (Pages 77–80)

A condition is an expression that can be evaluated as either true or false. When you use a condition in a WHERE clause, the results of the query contain those rows for which the condition is true. You can create simple conditions and compound conditions using the BETWEEN, LIKE, IN, EXISTS, ALL, and ANY operators, as described in the following sections.

Simple Conditions (Pages 77–78)

A simple condition has the form: column name, comparison operator, and then either another column name or a value. The available comparison operators are = (equal to), < (less than), > (greater than), <= (less than or equal to), >= (greater than or equal to), and < > (not equal to).

The following WHERE clause uses a condition to select rows where the balance is greater than the credit limit:

```
WHERE BALANCE > CREDIT_LIMIT
```

Compound Conditions (Pages 78–80)

Compound conditions are formed by connecting two or more simple conditions using the AND, OR, and NOT operators. When simple conditions are connected by the AND operator, all of the simple conditions must be true in order for the compound condition to be true. When simple conditions are connected by the OR operator, the compound condition will be true whenever any one of the simple conditions is true. Preceding a condition by the NOT operator reverses the truth of the original condition.

The following WHERE clause is true if the warehouse number is equal to 3 *or* the units on hand is greater than 100, *or* both:

```
WHERE (WAREHOUSE = '3') OR (ON_HAND > 100)
```

The following WHERE clause is true if the warehouse number is equal to 3 and the units on hand is greater than 100:

```
WHERE (WAREHOUSE = '3') AND (ON_HAND > 100)
```

The following WHERE clause is true if the warehouse number is not equal to 3:

```
WHERE NOT (WAREHOUSE = '3')
```

BETWEEN Conditions (Pages 80–81)

You can use the BETWEEN operator to determine if a value is within a range of values.

The following WHERE clause is true if the balance is between 2,000 and 5,000:

```
WHERE BALANCE BETWEEN 2000 AND 5000
```

LIKE Conditions (Pages 83–84)

LIKE conditions use wildcards to select rows. Use the percent (%) wildcard to represent any collection of characters. The condition LIKE '%Central%' will be true for data consisting of any character or characters, followed by the letters "Central," followed by any other character or characters. Another wildcard symbol is the underscore (_), which represents any individual character. For example, "T_m" represents the letter "T," followed by any single character, followed by the letter "m," and would be true for a collection of characters such as Tim, Tom, or T3m.

The following WHERE clause is true if the value in the STREET column is Central, Centralia, or any other value that contains "Central":

```
WHERE STREET LIKE '%Central%'
```

IN Conditions (Pages 84–85, 110)

You can use IN to determine whether a value is in some specific collection of values.

The following WHERE clause is true if the credit limit is 5,000, 10,000, or 15,000:

```
WHERE CREDIT_LIMIT IN (5000, 10000, 15000)
```

The following WHERE clause is true if the part number is in the collection of part numbers associated with order number 21610:

```
WHERE PART_NUM IN
(SELECT PART_NUM
FROM ORDER_LINE
WHERE ORDER_NUM = '21610')
```

EXISTS Conditions (Pages 110–112)

You can use EXISTS to determine whether the results of a subquery contain at least one row.

The following WHERE clause is true if the results of the subquery contain at least one row, that is, there is at least one order line with the desired order number and on which the part number is DR93:

```
WHERE EXISTS
(SELECT *
FROM ORDER_LINE
WHERE ORDERS.ORDER_NUM = ORDER_LINE.ORDER_NUM
AND PART_NUM = 'DR93')
```

ALL and ANY (Pages 125–128)

You can use ALL or ANY with subqueries. If you precede the subquery by ALL, the condition is true only if it is satisfied for all values produced by the subquery. If you precede the subquery by ANY, the condition is true if it is satisfied for any value (one or more) produced by the subquery.

The following WHERE clause is true if the balance is greater than every balance contained in the results of the subquery:

```
WHERE BALANCE > ALL
(SELECT BALANCE
FROM CUSTOMER
WHERE REP_NUM = '65')
```

The following WHERE clause is true if the balance is greater than at least one balance contained in the results of the subquery:

```
WHERE BALANCE > ANY
(SELECT BALANCE
FROM CUSTOMER
WHERE REP_NUM = '65')
```

■ CREATE INDEX (Pages 176–177)

Use the CREATE INDEX command to create an index for a table. Figure A-3 describes the CREATE INDEX command.

FIGURE A-3   CREATE INDEX command

Clause	Description	Required?
CREATE INDEX *index name*	Indicates the name of the index.	Yes
ON *table name*	Indicates the table for which the index is to be created.	Yes
column list	Indicates the column or columns on which the index is to be based.	Yes

The following CREATE INDEX command creates an index named REPNAME for the REP table on the combination of the LAST_NAME and FIRST_NAME columns:

```
CREATE INDEX REPNAME ON REP(LAST_NAME, FIRST_NAME);
```

■ CREATE TABLE (Pages 31–33)

Use the CREATE TABLE command to define the structure of a new table. Figure A-4 describes the CREATE TABLE command.

FIGURE A-4   CREATE TABLE command

Clause	Description	Required?
CREATE TABLE *table name*	Indicates the name of the table to be created.	Yes
(column and data type list)	Indicates the columns that comprise the table along with their corresponding data types (see Data Types section).	Yes

The following CREATE TABLE command creates the REP table and its associated columns and data types. REP_NUM is the table's primary key.

```
CREATE TABLE REP
(REP_NUM CHAR(2) PRIMARY KEY,
LAST_NAME CHAR(15),
FIRST_NAME CHAR(15),
STREET CHAR(15),
CITY CHAR(15),
STATE CHAR(2),
ZIP CHAR(5),
COMMISSION DECIMAL(7,2),
RATE DECIMAL(3,2) );
```

■ CREATE VIEW (Pages 160–169)

Use the CREATE VIEW command to create a view. Figure A-5 describes the CREATE VIEW command.

FIGURE A-5   CREATE VIEW command

Clause	Description	Required?
CREATE VIEW *view name* AS	Indicates the name of the view to be created.	Yes
query	Indicates the defining query for the view.	Yes

The following CREATE VIEW command creates a view named HOUSEWARES, which consists of the part number, part description, units on hand, and unit price for all rows in the PART table on which the ITEM class is HW:

```
CREATE VIEW HOUSEWARES AS
SELECT PART_NUM, PART_DESCRIPTION, ON_HAND, PRICE
FROM PART
WHERE CLASS = 'HW';
```

■ Data Types

Figure A-6 describes the data types that you can use in a CREATE TABLE command.

FIGURE A-6   Data types

Data Type	Description
CHAR(n)	Stores a character string n characters long. You use the CHAR type for columns that contain letters and special characters and for columns containing numbers that will not be used in any arithmetic operations. Because neither sales rep numbers nor customer numbers will be used in any arithmetic operations, for example, the REP_NUM and CUSTOMER_NUM columns are both assigned the CHAR data type.
DATE	Stores date data. The specific format in which dates are stored varies from one SQL implementation to another. In Oracle, dates are enclosed in single quotation marks and have the form DD-MON-YYYY (for example, '15-OCT-2007' is October 15, 2007). In Access, dates are enclosed in number signs and have the form MM/DD/YYYY (for example, #10/15/2007# is October 15, 2007). In MySQL, dates are enclosed in single quotation marks and have the form YYYY-MM-DD (for example, '2007-10-15' is October 15, 2007).
DECIMAL(p,q)	Stores a decimal number p digits long with q of these digits being decimal places to the right of the decimal point. For example, the data type DECIMAL(5,2) represents a number with three places to the left and two places to the right of the decimal (e.g., 100.00). You can use the contents of DECIMAL columns for arithmetic. (*Note:* The specific meaning of DECIMAL varies from one SQL implementation to another. In some implementations, the decimal point counts as one of the places, and in other implementations it does not. Likewise, in some implementations a minus sign counts as one of the places, but in others it does not.)
INTEGER	Stores integers, which are numbers without a decimal part. The valid range is −2147483648 to 2147483647. You can use the contents of INTEGER columns for arithmetic.
SMALLINT	Stores integers, but uses less space than the INTEGER data type. The valid range is −32768 to 32767. SMALLINT is a better choice than INTEGER when you are certain that the column will store numbers within the indicated range. You can use the contents of SMALLINT columns for arithmetic.

■ DELETE Rows (Page 146)

Use the DELETE command to delete one or more rows from a table. Figure A-7 describes the DELETE command.

FIGURE A-7 DELETE command

Clause	Description	Required?
DELETE *table name*	Indicates the table from which the row or rows are to be deleted.	Yes
WHERE *condition*	Indicates a condition. Those rows for which the condition is true will be retrieved and deleted.	No (If you omit the WHERE clause, all rows will be deleted.)

The following DELETE command deletes any row from the LEVEL1_CUSTOMER table on which the customer number is 895:

```
DELETE LEVEL1_CUSTOMER
WHERE CUSTOMER_NUM = '895';
```

■ DROP INDEX (Pages 177–178)

Use the DROP INDEX command to delete an index, as shown in Figure A-8.

FIGURE A-8 DROP INDEX command

Clause	Description	Required?
DROP INDEX *index name*	Indicates the name of the index to be dropped.	Yes

The following DROP INDEX command deletes the index named CREDNAME:

```
DROP INDEX CREDNAME;
```

■ DROP TABLE (Pages 152–153)

Use the DROP TABLE command to delete a table, as shown in Figure A-9.

FIGURE A-9 DROP TABLE command

Clause	Description	Required?
DROP TABLE *table name*	Indicates name of table to be dropped.	Yes

The following DROP TABLE command deletes the table named LEVEL1_CUSTOMER:

```
DROP TABLE LEVEL1_CUSTOMER;
```

■ DROP VIEW (Page 170)

Use the DROP VIEW command to delete a view, as shown in Figure A-10.

FIGURE A-10   DROP VIEW command

Clause	Description	Required?
DROP VIEW *view name*	Indicates the name of the view to be dropped.	Yes

The following DROP VIEW command deletes the view named HSEWRES:

```
DROP VIEW HSEWRES;
```

■ GRANT (Pages 170–173)

Use the GRANT command to grant privileges to a user. Figure A-11 describes the GRANT command.

FIGURE A-11   GRANT command

Clause	Description	Required?
GRANT *privilege*	Indicates the type of privilege(s) to be granted.	Yes
ON *database object*	Indicates the database object(s) to which the privilege(s) pertain.	Yes
TO *user name*	Indicates the user(s) to whom the privilege(s) are to be granted. To grant the privilege(s) to all users, use the TO PUBLIC clause.	Yes

The following GRANT command grants the user named Johnson the privilege of selecting rows from the REP table:

```
GRANT SELECT
ON REP
TO Johnson;
```

■ INSERT INTO (Query) (Page 141)

Use the INSERT INTO command with a query to insert the rows retrieved by a query into a table. As shown in Figure A-12, you must indicate the name of the table into which the row(s) will be inserted and the query whose results will be inserted into the named table.

FIGURE A-12   INSERT INTO (query) command

Clause	Description	Required?
INSERT INTO *table name*	Indicates the name of the table into which the row(s) will be inserted.	Yes
query	Indicates the query whose results will be inserted into the table.	Yes

The following INSERT INTO command inserts rows selected by a query into the LEVEL1_CUSTOMER table:

```
INSERT INTO LEVEL1_CUSTOMER
SELECT CUSTOMER_NUM, CUSTOMER_NAME, BALANCE, CREDIT_LIMIT, REP_NUM
FROM CUSTOMER
WHERE CREDIT_LIMIT = 7500;
```

■ INSERT INTO (Values) (Pages 44–49, 143–144)

Use the INSERT INTO command and the VALUES clause to insert a row into a table by specifying the values for each of the columns. As shown in Figure A-13, you must indicate the table into which to insert the values, and then list the values to insert in parentheses.

FIGURE A-13   INSERT INTO (values) command

Clause	Description	Required?
INSERT INTO *table name*	Indicates the name of the table into which the row will be inserted.	Yes
VALUES *(values list)*	Indicates the values for each of the columns on the new row.	Yes

The following INSERT INTO command inserts the values shown in parentheses as a new row in the REP table:

```
INSERT INTO REP
VALUES
('20','Kaiser','Valerie','624 Randall','Grove','FL','33321',20542.50,0.05);
```

■ Integrity (Pages 181–184)

You can use the ALTER TABLE command with an appropriate ADD CHECK, ADD PRIMARY KEY, or ADD FOREIGN KEY clause to specify integrity. Figure A-14 describes the ALTER TABLE command for specifying integrity.

FIGURE A-14   Integrity options

Clause	Description	Required?
ALTER TABLE *table name*	Indicates the table for which integrity is being specified.	Yes
integrity clause	ADD CHECK, ADD PRIMARY KEY, or ADD FOREIGN KEY	Yes

The following ALTER TABLE command changes the PART table so that the only legal values for the CLASS column are AP, HW, and SG:

```
ALTER TABLE PART
ADD CHECK (CLASS IN ('AP','HW','SG') );
```

The following ALTER TABLE command changes the REP table so that the REP_NUM column is the table's primary key:

```
ALTER TABLE REP
ADD PRIMARY KEY(REP_NUM);
```

The following ALTER TABLE command changes the CUSTOMER table so that the REP_NUM column in the CUSTOMER table is a foreign key referencing the primary key of the REP table:

```
ALTER TABLE CUSTOMER
ADD FOREIGN KEY(REP_NUM) REFERENCES REP;
```

■ REVOKE (Page 173)

Use the REVOKE command to revoke privileges from a user. Figure A-15 describes the REVOKE command.

FIGURE A-15   REVOKE command

Clause	Description	Required?
REVOKE *privilege*	Indicates the type of privilege(s) to be revoked.	Yes
ON *database object*	Indicates the database object(s) to which the privilege pertains.	Yes
FROM *user name*	Indicates the user name(s) from whom the privilege(s) are to be revoked.	Yes

The following REVOKE command revokes the SELECT privilege for the REP table from the user named Johnson:

```
REVOKE SELECT
ON REP
FROM Johnson;
```

■ ROLLBACK (Pages 144–145, 147)

Use the ROLLBACK command to reverse (undo) all updates since the execution of the previous COMMIT command. If no COMMIT command has been executed, the command will undo all changes made during the current work session. Figure A-16 describes the ROLLBACK command.

FIGURE A-16   ROLLBACK command

Clause	Description	Required?
ROLLBACK	Indicates that a rollback is to be performed.	Yes

The following command reverses all updates made since the time of the last COMMIT command:

```
ROLLBACK;
```

■ SELECT (Pages 75–98, 106–109)

Use the SELECT command to retrieve data from a table or from multiple tables. Figure A-17 describes the SELECT command.

Clause	Description	Required?
SELECT *column or expression list*	Indicates the column(s) and/or expression(s) to be retrieved.	Yes
FROM *table list*	Indicates the table(s) required for the query.	Yes
WHERE *condition*	Indicates one or more conditions. Only the rows for which the condition(s) are true will be retrieved.	No (If you omit the WHERE clause, all rows will be retrieved.)
GROUP BY *column list*	Indicates column(s) on which rows are to be grouped.	No (If you omit the GROUP BY clause, no grouping will occur.)
HAVING *condition involving groups*	Indicates a condition for groups. Only groups for which the condition is true will be included in query results. Use the HAVING clause only if the query output is grouped.	No (If you omit the HAVING clause, all groups will be included.)
ORDER BY *column or expression list*	Indicates column(s) on which the query output is to be sorted.	No (If you omit the ORDER BY clause, no sorting will occur.)

The following SELECT command joins the ORDERS and ORDER_LINE tables. The command selects the customer number, order number, order date, and the sum of the product of the number ordered and unit price, renamed as ORDER_TOTAL. Records are grouped by order number, customer number, and date. Only groups on which the order total is greater than 1,000 are included. Groups are ordered by order number.

```
SELECT CUSTOMER_NUM, ORDERS.ORDER_NUM, ORDER_DATE,
SUM(NUM_ORDERED * QUOTED_PRICE) AS ORDER_TOTAL
FROM ORDERS, ORDER_LINE
WHERE ORDERS.ORDER_NUM = ORDER_LINE.ORDER_NUM
GROUP BY ORDERS.ORDER_NUM, CUSTOMER_NUM, ORDER_DATE
HAVING SUM(NUM_ORDERED * QUOTED_PRICE) > 1000
ORDER BY ORDERS.ORDER_NUM;
```

■ Subqueries (Pages 91–93, 112–113)

You can use one query within another. The inner query is called a subquery and it is evaluated first. The outer query is evaluated next, producing the part description for each part whose part number is in the list.

The following command contains a subquery that produces a list of part numbers included in order number 21610:

```
SELECT PART_DESCRIPTION
FROM PART
WHERE PART_NUM IN
(SELECT PART_NUM
FROM ORDER_LINE
WHERE ORDER_NUM = '21610');
```

■ UNION, INTERSECT, and MINUS (Pages 121–125)

Connecting two SELECT commands with the UNION operator produces all the rows that would be in the results of the first query, the second query, or both queries. Connecting two SELECT commands with the INTERSECT operator produces all the rows that would be in the results of both queries. Connecting two SELECT commands with the MINUS operator produces all the rows that would be in the results of the first query, but *not* in the results of the second query. Figure A-18 describes the UNION, INTERSECT, and MINUS operators.

FIGURE A-18   UNION, INTERSECT, and MINUS operators

Operator	Description
UNION	Produces all the rows that would be in the results of the first query, the second query, or both queries.
INTERSECT	Produces all the rows that would be in the results of both queries.
MINUS	Produces all the rows that would be in the results of the first query but not in the results of the second query.

The following query displays the customer number and customer name of all customers that are represented by sales rep 65, *or* that have orders, *or* both:

```
SELECT CUSTOMER_NUM, CUSTOMER_NAME
FROM CUSTOMER
WHERE REP_NUM = '65'
UNION
SELECT CUSTOMER.CUSTOMER_NUM, CUSTOMER_NAME
FROM CUSTOMER, ORDERS
WHERE CUSTOMER.CUSTOMER_NUM = ORDERS.CUSTOMER_NUM;
```

The following query displays the customer number and customer name of all customers that are represented by sales rep 65 *and* that have orders:

```
SELECT CUSTOMER_NUM, CUSTOMER_NAME
FROM CUSTOMER
WHERE REP_NUM = '65'
INTERSECT
SELECT CUSTOMER.CUSTOMER_NUM, CUSTOMER_NAME
FROM CUSTOMER, ORDERS
WHERE CUSTOMER.CUSTOMER_NUM = ORDERS.CUSTOMER_NUM;
```

The following query displays the customer number and customer name of all customers that are represented by sales rep 65 but that do *not* have orders:

```
SELECT CUSTOMER_NUM, CUSTOMER_NAME
FROM CUSTOMER
WHERE REP_NUM = '65'
MINUS
SELECT CUSTOMER.CUSTOMER_NUM, CUSTOMER_NAME
FROM CUSTOMER, ORDERS
WHERE CUSTOMER.CUSTOMER_NUM = ORDERS.CUSTOMER_NUM;
```

■ UPDATE (Pages 141–143)

Use the UPDATE command to change the contents of one or more rows in a table. Figure A-19 describes the UPDATE command.

FIGURE A-19   UPDATE command

Clause	Description	Required?
UPDATE *table name*	Indicates the table whose contents will be changed.	Yes
SET *column = expression*	Indicates the column to be changed, along with an expression that provides the new value.	Yes
WHERE *condition*	Indicates a condition. The change will occur only on those rows for which the condition is true.	No (If you omit the WHERE clause, all rows will be updated.)

The following UPDATE command changes the customer name on the row in LEVEL1_CUSTOMER on which the customer number is 842 to All Season Sport:

```
UPDATE LEVEL1_CUSTOMER
SET CUSTOMER_NAME = 'All Season Sport'
WHERE CUSTOMER_NUM = '842';
```

APPENDIX

"How Do I" Reference

This appendix answers frequently asked questions about how to accomplish a variety of tasks using SQL. Use the second column to locate the correct section in Appendix A that answers your question.

How Do I	Review the Named Section(s) in Appendix A
Add columns to an existing table?	ALTER TABLE
Add rows?	INSERT INTO (Values)
Calculate a statistic (sum, average, maximum, minimum, or count)?	1. SELECT 2. Column or Expression List (SELECT Clause) (Use the appropriate function in the query.)
Change rows?	UPDATE
Create a data type for a column?	1. Data Types 2. CREATE TABLE
Create a table?	CREATE TABLE
Create a view?	CREATE VIEW
Create an index?	CREATE INDEX
Delete a table?	DROP TABLE
Delete a view?	DROP VIEW
Delete an index?	DROP INDEX
Delete rows?	DELETE ROWS
Drop a table?	DROP TABLE
Drop a view?	DROP VIEW
Drop an index?	DROP INDEX
Grant a privilege?	GRANT
Group data in a query?	SELECT (Use a GROUP BY clause.)
Insert rows?	INSERT INTO (Values)
Insert rows using a query?	INSERT INTO (Query)
Join tables?	Conditions (Include a WHERE clause to relate the tables.)
Make updates permanent?	COMMIT
Order query results?	SELECT (Use the ORDER BY clause.)
Prohibit nulls?	1. CREATE TABLE 2. ALTER TABLE (Include the NOT NULL clause in a CREATE TABLE or ALTER TABLE command.)
Remove a privilege?	REVOKE
Remove rows?	DELETE Rows
Retrieve all columns?	1. SELECT 2. Column or Expression List (SELECT Clause) (Type * in the SELECT clause.)

How Do I	Review the Named Section(s) in Appendix A
Retrieve all rows?	SELECT (Omit the WHERE clause.)
Retrieve only certain columns?	1. SELECT 2. Column or Expression List (SELECT Clause) (Type the list of columns in the SELECT clause.)
Revoke a privilege?	REVOKE
Select all columns?	1. SELECT 2. Column or Expression List (SELECT Clause) (Type * in the SELECT clause.)
Select all rows?	SELECT (Omit the WHERE clause.)
Select only certain columns?	1. SELECT 2. Column or Expression List (SELECT Clause) (Type the list of columns in the SELECT clause.)
Select only certain rows?	1. SELECT 2. Conditions (Use a WHERE clause.)
Sort query results?	SELECT (Use an ORDER BY clause.)
Specify a foreign key?	Integrity (Use the ADD FOREIGN KEY clause in an ALTER TABLE command.)
Specify a primary key?	Integrity (Use the ADD PRIMARY KEY clause in an ALTER TABLE command.)
Specify a privilege?	GRANT
Specify integrity?	Integrity (Use an ADD CHECK, ADD PRIMARY KEY, and/or ADD FOREIGN KEY clause in an ALTER TABLE command.)
Specify legal values?	Integrity (Use an ADD CHECK clause in an ALTER TABLE command.)
Undo updates?	ROLLBACK
Update rows?	UPDATE
Use a calculated field?	1. SELECT 2. Column or Expression List (SELECT Clause) (Enter a calculation in the query.)
Use a compound condition?	1. SELECT 2. Conditions (Use simple conditions connected by AND, OR, or NOT in a WHERE clause.)
Use a compound condition in a query?	Conditions

How Do I	Review the Named Section(s) in Appendix A
Use a condition in a query?	1. SELECT 2. Conditions (Use a WHERE clause.)
Use a subquery?	Subqueries
Use a wildcard?	1. SELECT 2. Conditions (Use LIKE and a wildcard in a WHERE clause.)
Use an alias?	Aliases (Enter an alias after the name of each table in the FROM clause.)
Use set operations (union, intersection, difference)?	UNION, INTERSECT, and MINUS (Connect two SELECT commands with UNION, INTERSECT, or MINUS.)

APPENDIX C

Answers to Odd-Numbered Review Questions

■ Chapter 1—Introduction to Premiere Products, Henry Books, and Alexamara Marina Group

Due to the nature of the material in Chapter 1, there are no Review Questions.

■ Chapter 2—An Introduction to SQL

1. A relation is a two-dimensional table in which the entries in the table are single-valued (each location in the table contains a single entry), each column has a distinct name (or attribute name), all values in a column are values of the same attribute, the order of the rows and columns is immaterial, and each row contains unique values.

3. Rows in a relation are also called records or tuples. Columns are also called fields or attributes.

5. The primary key of a table is the column or collection of columns that uniquely identifies a given row in the table. You usually identify primary keys by underlining the column name(s).

7. INTEGER, SMALLINT, DECIMAL, CHAR, and DATE

9. Use the INSERT command.

11. Use the UPDATE command.

13. Use the DESCRIBE command in Oracle SQL*Plus or Oracle SQL*Plus Worksheet. Use the Documenter tool in Access. In MySQL, use the SHOW COLUMNS command.

■ Chapter 3—Single-Table Queries

1. The basic form of the SELECT command is SELECT-FROM-WHERE. Specify the columns to be listed after the word SELECT (or type * to select all columns), and then specify the table name that contains these columns after the word FROM. Optionally, you can include condition(s) after the word WHERE.

3. You can form a compound condition by combining simple conditions and using the operators AND, OR, or NOT.

5. Use arithmetic operators and write the computation in place of a column name. You can assign a name to the computation by following the computation with the word AS and then the desired name.

7. The percent (%) wildcard represents any collection of characters. The underscore (_) wildcard represents any single character. In Access, the asterisk (*) wildcard represents any collection of characters, and the question mark (?) wildcard represents any single character.

9. Use an ORDER BY clause.

11. To sort data in descending order, follow the sort key with the word DESC.

13. To avoid duplicates, precede the column name with the DISTINCT operator.

15. Use a GROUP BY clause.

17. Use an IS NULL operator in the WHERE clause.

■ Chapter 4—Multiple-Table Queries

1. Indicate in the SELECT clause all columns to display, list in the FROM clause all tables to join, and then include in the WHERE clause any conditions requiring values in matching columns to be equal.

3. IN and EXISTS

5. An alias is an alternate name for a table. To specify an alias in SQL, follow the name of the table with the name of the alias. You use the alias just like a table name throughout the SQL command.

7. Use the UNION, INTERSECT, and MINUS operators to create a union, intersection, and difference of two tables. To perform any of these operations, the tables must be union compatible.

9. If the ALL operator precedes a subquery, the condition is true only if it is satisfied by all values produced by the subquery.

11. In an inner join, only matching rows from both tables are included. You can use the INNER JOIN clause to perform an inner join.

13. In a right outer join, all rows from the table on the right will be included regardless of whether they match rows from the table on the left. Rows from the table on the left will be included only if they match. You can use the RIGHT JOIN clause to perform a right outer join.

■ Chapter 5—Updating Data

1. CREATE TABLE

3. Use the INSERT command with a SELECT clause.

5. COMMIT

7. Before beginning the updates for a transaction, commit any previous updates by executing the COMMIT command. Complete the updates for the transaction. If any update cannot be completed, execute the ROLLBACK command and discontinue the updates for the current transaction. If you can complete all updates successfully, execute the COMMIT command after completing the final update.

9. The clause is SET followed by the column name, followed by an equals sign (=) and the word NULL.

11. The ALTER TABLE command with the MODIFY clause.

■ Chapter 6—Database Administration

1. A view contains data that is derived from existing base tables when users attempt to access the view.

3. A defining query is the portion of the CREATE VIEW command that describes the data to include in the view.

5. Views provide data independence, allow database access control, and simplify the database structure for users.

7. DROP VIEW

9. REVOKE

11. Use the CREATE INDEX command to create an index. Use the CREATE UNIQUE INDEX command to create a unique index. A unique index allows only unique values in the column (or columns) on which the index is created.

13. The DBMS

15. SELECT

17. Integrity constraints are rules that the data in the database must follow to ensure that only legal values are accepted in specified columns, or that primary and foreign key values match between tables.

19. Primary key constraints are usually specified when you create a table. You can specify them later using the PRIMARY KEY clause of the ALTER TABLE command.

■ Chapter 7—Reports

1. Use the UPPER function to convert letters to uppercase. Use the LOWER function to convert letters to lowercase.

3. To add months to a date, use the ADD_MONTHS function (Oracle), the ADDDATE function (MySQL), or the DATEADD function (Access). To add days to a date, simply add the desired number of days to a date (Oracle), use the ADDDATE function (MySQL), or use the DATEADD function (Access). To find the number of days between two dates, subtract the earlier date from the later date.

5. Separate the column names with two vertical lines (||) in the SELECT clause.

7. Execute a query that selects the data from a table or previously created view that you want to use in the report.

9. Type a single vertical line (|) at the point where you want to break the heading.

11. TTITLE (top), BTITLE (bottom)

13. SET PAGESIZE

15. Use the COMPUTE command and an appropriate statistical function. In order to do so, you must also use the BREAK command.

17. SPOOL

■ Chapter 8—Embedded SQL

1. In the procedural code

3. Use the %TYPE attribute.

5. Precede the name with an ampersand (&).

7. Use the INTO clause in the SELECT command.

9. Use the command to define a cursor.

11. FETCH

13. FOR UPDATE OF; to update data, include the WHERE CURRENT OF [cursor name] clause in the UPDATE command.

15. Create the command in a string variable. To run the command stored in the string variable, use the DoCmd.RunSQL command.

17. Use the MoveNext command.

INDEX

Access (Microsoft)
   & symbol, column concatenation, 200
   ALTER TABLE command, 151
   changing column names in, 163
   creating views in, 161
   currency field types in, 140
   DATE function, 198
   DATEADD function, 197, 198, 220
   Documenter tool, 63, 179
   foreign keys and referential integrity, 182
   indexes, dropping, 178
   query results, viewing, 74
   and PL/SQL programs, 231
   primary key, specifying, 182
   printing exercises, commands, 67
   programs, using SQL in, 249–255
   rollback method in, 147
   SQL queries in, 34, 49–55, 57, 255
   UCASE, LCASE functions, 195
   validation rules, specifying, 184
   views, creating, 161
   wildcard symbols, 84
access control to databases, 170–173
ADD clause, 153
ADD FOREIGN KEY, ADD PRIMARY KEY
   clauses, 182, 185
ADDDATE function, 220
adding
   new rows to tables, 143–144, 166
   title to reports, 209–210
ADD_MONTHS function, 196–197, 220
administration, database. *See* database administration
aggregate functions, 87, 99
Alexamara Marina Group
   database described, 2, 16–20
   exercises, 23, 70–72, 103, 137, 157–158,
      190–191, 227–228, 259
   organization described, 21

aliases
   described, using, 113–114
   using in joins, 118–119
ALL operator
   described, using, 125–128
   using with subqueries to produce columns of
      numbers, 125–126
ALTER TABLE command, 148–149, 153, 181, 184
ampersand (&) symbol, column concatenation, 200
AND operator, 78–79
ANY operator, 125–127
arguments, including in commands, 249–250
asterisks
   in Access, 84
   and SELECT command, 76
   in SQL commands, 49
attributes
   database, described, 28–31
   %TYPE attribute, 234

BALANCE condition, 93
base tables and table views, 160
BEGIN command, 233
BETWEEN operator, 80–81, 99
BREAK command, 211, 220
BTITLE command, 209, 217
buffers, command, 36

C/CHR/CHAR command, 38
Cartesian Product, 131, 133
case-sensitivity, SQL commands, 30, 78
catalog. *See* system catalog
changing
   data in tables, 141–143
   saving, 144–145
   single row in table, 237
   table structure, 148–152

values in columns, 55–56
values in columns to nulls, 147–148
CHAR data type
   described, 42
   quotation marks ("") and, 44
character functions, 194–195
CHECK clause in ALTER TABLE command, 183, 185
clauses
   See also specific clauses
   SQL query (fig.), 98
CLEAR BREAK command, 217, 220
CLEAR COLUMNS command, 206, 217
CLEAR COMPUTE command, 220
CLOSE command, 240, 243
collections in Access, 84
COLUMN command, 206, 220
column names
   changing in Access, 163
   qualifying, 106, 107, 132
columns
   changing values in, 55–56
   computed, 81–82, 99
   concatenating, 199–201
   in databases, 28
   headings in reports, 205–209
   retrieving specific, 75–85
combining table rows with product operation, 131
command buffers, 36
commands
   See also specific commands
   Access. *See* Access
   MySQL. *See* MySQL
   Oracle SQL *Plus. *See* SQL *Plus
   PL/SQL, 233
   report-formatting (fig.), 218
   undoing, 144–145
COMMIT command, 144–145, 153
comparison operators in SQL commands (fig.), 77–78
compound conditions, 78–80, 98
COMPUTE command, 212–213
computed columns, 81–82, 99
CONCAT function, 220

concatenating
   columns, 199–201
   values in character columns, 220
conditions
   in joins, 117
   simple, 77, 98
constraints, integrity, 181
constructing simple queries, 75–85
correcting errors
   in data, 56
   in SQL commands, 36–40
correlated subqueries, 111
COUNT function, 87–88
CREATE INDEX command, 177, 185
CREATE TABLE command, 31–33, 35–37, 140–141
   for CUSTOMER table (fig.), 58
   with errors (fig.), 39
   and table structure, 62–65
CREATE VIEW command, 161, 164, 184, 203
creating
   database tables, 58–62
   databases, 31–33
   defining queries, 160–161
   indexes, 176–177
   new tables from existing tables, 140–141
   scripts, 202
   tables, 35–36, 153
   text files containing SQL commands, 57
   views definition, 161
CURRENCY field type, 140
cursors
   complete program using, 243–244
   complex, 244–246
   described, using, 240–248
customer orders, database information for, 4

data
   changing in tables, 141–143
   creating for report, 204–205
   grouping data in databases, 93–95
   independence, and views, 165
   loading tables with, 44–49
   sorting in databases, 85–87
   updating, generally, 140
   viewing table, 49–55

data dictionary, 178

data types
in Access, 35
common (fig.), 42
described, 31, 65
and %TYPE attribute, 234

database administration process, 160

database administrators, 160, 173

database management systems. *See* DBMS (database management system)

database tables, creating, 58–62

databases
correcting errors in, 55–56
creating, 31–33
described, 2
relational. *See* relational databases
sorting data in, 85–87

DATE data type, 42

dates, SQL functions, 196–199

DBMS (database management system)
indexes in, 173–174
integrity support, 181
Microsoft Access. *See* Access
optimizers described, 246
SQL use, 26
support for views, 160
system catalog, 178–181

Debug.Print, 251

DECIMAL data type, 35, 42

defining queries, 160–161, 165

DELETE command, 56, 65, 146, 153, 238, 250, 255

deleting
See also dropping
indexes, 177–178
rows, 145–146, 238, 250
tables, 152
views, 170

DESCRIBE command, 62, 152

difference of tables, 122

Dim statement, 250

DISTINCT operator, 89–91, 99, 166–167

DoCmd.RunSQL command, 250

Documenter tool (Access), 63, 179

DROP INDEX command, 177

DROP TABLE command, 41, 65, 152–153, 185

dropping
See also deleting
database tables, 41
tables, 152
views, 170

EDIT command, 45

editing commands
commands in MySQL, 39, 47–48
MySQL (fig.), 39
SQL, 36–40
SQL *Plus, 36, 37

embedded SQL
error handling, 249
introduction to, 230–231
optimizers described, 246
running in Access programs, 249–255

END command, 233

entities, database, 28–31, 65

error handling, 249

errors
database, correcting, 55–56
in SQL commands, 36–40

EXCEPTION command, 249, 255

EXISTS operator
described, 132
and joins, 109
retrieving data from multiple tables, 110–112

FETCH command, 240–243, 245–246, 255

fields
CURRENCY type, 140
in databases, 28

files
script, 57
sending reports to, 214–216

FLOOR function, 196, 219

FOR UPDATE OF clause, 247–248, 255

FORMAT clause, 207

free-format commands, 32

FROM clause
in inner joins, 128–129
in joins, 109, 115–116
and SELECT command, 75

full outer joins, 129

functions
    See also specific functions
    described, using, 87–91, 219
    using, 194–199

GRANT command, 160, 171–173, 185
GROUP BY clause
    and calculating sums, 194
    described, using, 94–95, 99
grouping data
    in databases, 93–95
    in reports, 211–212

HAVING clause, 95–97, 99
HEADING clause, 206, 220
Henry Books
    database described, 2, 9–15
    exercises, 22, 68–70, 101–102, 136, 156,
        189–190, 225–227, 258
    organization described, 9, 21
    sample data (figs.), 10, 11

implementing nulls, 43
IN clause, 84–85
IN operator
    described, using, 84–85, 132
    and joins, 109
    retrieving data from multiple tables, 110
indexes, using, 173–178
inner joins, 128–129
INSERT command
    in CUSTOMER table (fig.), 58–60
    described, using, 44–49, 65, 153
    modifying, 45–46
    updating data with, 143–144
inserting
    data into tables, 140–141
    rows into tables, 236, 253
INTEGER data type, 42
integrity rules in SQL, 181–185
INTERSECT command, 123–124, 132
intersections, using, 121, 124–125
INTO clause and FETCH, 241, 245–246
IS NULL operator, 99

joining
    multiple tables, 119–121
    table to itself (self-join), 115–116
    tables generally, 106
    two tables, 106–109
joins
    inner, and outer, 129–130
    retrieving single row from, 235
    views of, 167–170

key, primary. *See* primary key
keys, sort, 86

L command, 38
left and right outer joins, 129–130, 133
LIKE operator, 83–84, 99
line breaks, adding, 211, 220
loading tables with data, 44–49
LOWER function, 195

Microsoft
    Access. *See* Access
    DBMS described, 34–35
minus, of tables, 122
MINUS operator, 124–125, 132
mistakes. *See* errors
MODIFY clause, using with ALTER TABLE, 151
MySQL
    ADDDATE function, 197, 198
    command syntax, execution, 52
    commands, editing, 36, 47
    COMMIT and ROLLBACK commands, 145
    CONCAT function, 200
    CREATE TABLE command with errors (fig.), 39
    CURDATE function, 198
    dropping indexes, 178
    editing commands in, 47–48
    query results display, 74
    and PL/SQL programs, 231
    printing exercises, commands, 67
    program described, 35–36
    rollback method in, 147
    script files in, 57
    SHOW COLUMNS command for REP table
        (fig.), 64

SHOW TABLES command, 179
statement history, 39
values, concatenating in character columns, 220
viewing query results, 74
viewing table data, 52–54
views and, 161
wildcards in, 83

names
    and aliases, 113–114
    changing column names in Access, 163
    computed columns, 82
    qualifying column, 106, 107, 132
    table, columns restrictions, 32
    variables in PL/SQL, 233
nesting
    queries, 91–93
    SELECT commands, 74
    subqueries, 112–113
NO_DATA_FOUND error, 249
NOT NULL clause, 43, 65
NOT operator, 78–79
Notepad text editor, 45
NULL operator, using, 99
nulls
    changing values in columns to, 147–148, 153
    in conditions, 97
    described, using, 42–43, 65
    INSERT command with, 48–49
    in SUM, AVG, MAX, MIN functions, 89
number functions, using, 195–199

ON clause, 213
one-to-many relationships, 28
OPEN command, 240
open-source programs, 35–36
operations
    intersection, 121, 124–125
    product, 131
    set, 121–125
    union, 121–123
operators
    See also specific operators
    comparison, 77–78
    SQL query (fig.), 98

optimizing
    data access with optimizers, 246
    data retrieval with indexes, 176
    queries, 113
OR operator, 78–79
Oracle
    PL/SQL. See PL/SQL
    SQL *Plus. See SQL *Plus
order, database terminology, 6
ORDER BY clause, 85–87, 99, 106
orders
    entering with transactions, 145–146
    Premiere Products sample (fig.), 3
ORDERS table, 5–6
outer joins, 129–130, 133

parentheses ( )
    around conditions, 80
    and embedded SQL, 251
percent sign (%)
    %TYPE attribute, 234
    wildcards, 83–84, 99
pipe character (|) and column concatenation, 199–201, 220
PL/SQL (Oracle)
    and multiple-row SELECT, 239–249
    programming language, 230–231
    programs, writing, 232
pointers. See cursors
Premiere Products
    database described, 2–9
    exercises, 21, 67, 100–101, 134–135, 155, 187–188, 222–224, 257
    organization described, 2, 21
    sample data for (fig.), 5
    sample order (fig.), 3
primary key
    in databases, 31, 65
    nulls and, 43
    self-join on, 118–121
    and updating joins, 170
    using self-join on, 118–121
printing
    exercises, SQL commands, 67
    reports, 214–216

privileges
    revoking users', 173
    to view system catalog data, 179
product operations, 131
programming languages
    embedded SQL and, 230
    procedural and nonprocedural, 230–231
    SQL. *See* SQL
prompt variables, 231–232, 236

• •

qualifying column names, 106, 107, 132
queries
    constructing simple, 75–85
    data retrieval and cursors, 244–246
    defining, creating, 160–161
    multiple tables, 106–117
    nesting, 91–93
    running report, 202–203
    saving SQL commands, 49–55
    SQL clauses and operators (fig.), 98
    using indexes with, 173
    using SELECT command, 49–55
    without asterisks, 88
question mark (?)
    in Access, 84
    and embedded SQL, 251
    wildcard, 99
quotation marks ("")
    for CHAR data type, 44
    and prompt variables, 232

• •

records in databases, 29
REFERENCES clause, 182
relational databases
    described, 26, 30, 65
    entities, attributes, relationships, 28–31
relationships
    between books, author, and branches data (fig.),
        13–14
    in databases, 2, 28–31
reports
    changing column headings, formats, 205–209
    creating with scripts, 201–202
    formatting commands (fig.), 218
    grouping data in, 211–212
    including totals and subtotals, 212–214

sending to a file, 214–216
    titles, adding to, 209–210
restricting
    rows when joining multiple tables, 118–121
    tables with WHERE clause, 179–180
retrieving
    data wit indexes, 174
    single row and column, 232–234
    single row from join, 235
    specific columns, rows, 75–85
reversing changes with ROLLBACK command,
    144–145
review questions
    database administration, 186
    embedded SQL, 256
    multiple-table queries, 134
    reports, 221–222
    single-table queries, 100
    SQL introduction, 66
    updating data, 154
REVOKE command, 160, 171, 185
right and left outer joins, 129–130, 133
ROLLBACK command, 140, 144–145
rollbacks, 140, 144–145, 147
ROUND function, 195–196, 219
row-and-column subsets, 163, 166–167
rows
    in databases, 29
    deleting, 145–146, 250
    finding, 253–255
    grouping, 93–95
    multiple-row SELECT, 239–249
    restricting with conditions, 108
    retrieving specific, 75–85
    updating, 252–253
RTRIM function, 220
RUN command, 46
running
    queries for generating reports, 202–203
    reports, 214–216
    SQL commands, 33–40

• •

sales representatives, database information about, 3–4
saving
    data, and COMMIT command, 144–145
    SQL commands, 56–57

script files, 57
scripts, 201–202, 216
security, and access control, 170–173
SELECT clause, 75–76
SELECT command
    described, 65, 98
    in embedded SQL, 255
    multiple-row SELECT, 239–249
    nesting, 74
    outputting as report, 194
    producing report with, 204
    running queries on table data, 49–55
    using for joins, 113, 119–121
    viewing table data, 49–55
self-joins
    described, using, 115–116
    incorrect (fig.), 117
    using on a primary key, 118–121
semicolons (;)
    in MySQL, 52
    in SQL *Plus, 50
SET clause and changing values in columns, 153
SET FEEDBACK OFF command, 211, 220
SET LINESIZE command, 51, 74, 210
set operations, using, 121–125
SET PAGESIZE command, 210
SET PAUSE ON command, 214–216, 220
SET SERVEROUTPUT ON
    command, 233
SHOW COLUMNS command, 64, 65
simple conditions, 77, 98
SKIP clause, 211, 220
slashes (/)
    in Oracle, 45
    use in scripts, 202, 234
SMALLINT data type, 42
sorting data in databases, 85–87
SPOOL, SPOOL OFF command, 215
SQL (Structured Query Language)
    case-sensitivity of, 30
    commands. *See* SQL commands
    embedded, 230–231, 246, 249–255
    integrity rules, using, 181–184
    MySQL. *See* MySQL
    programming language described, 26
    security mechanisms in, 171

    special operations in, 128–132
    SQL *Plus. *See* SQL *Plus
SQL commands
    See also specific commands
    case-sensitivity of, 30
    editing, 36–40
    embedding in other languages, 230–231
    free-format, 32
    running, 33–40
    saving, 49–55
.sql files, 202
SQL *Plus (Oracle)
    characters per line, changing, 74
    command syntax, execution, 33, 50
    command errors, correcting (fig.), 38
    DESCRIBE command, 62
    editing commands, 45
    script files, 57, 202
    scripts, creating in, 57
    editing commands, 36–37
    joins, left and right outer, 130
    modifying commands in, 45
    and PL/SQL programs, 230–231
    printing exercises, commands, 67
    printing reports, 215
    program described, 33–34
    Worksheet described, 33–34
SQL> prompt (SQL *Plus), 33, 46
starting
    Access Documenter, 63
    editing process, 40
statistics, views involving, 170
string variables, 250
strSQL string variable, 250
subqueries
    correlated, 111
    described, using, 92–93, 99, 132
    nesting, 112–113
subtotals, including in reports, 212–214
SUM function, 88–89
SUM(BALANCE) function, 194
support, integrity, 181
SYSDATE, 198
SYSTABLES table, 185
system catalog described, using, 178–181, 185

table structure, changing, 148–152

tables
  base, 160
  changing data in, 141–143
  correcting, 41
  creating, 58–62, 140–141
  defining structure of, 62–65
  deleting rows from, 145–146, 238–239
  dropping, 41
  inserting nulls into, 48–49
  inserting row into, 236
  intersections, 121, 124–125
  joining multiple, 119–121
  joining two, 106–109
  loading with data, 44–49
  querying multiple, 106–117
  self-joins, 115–116
  structure, changing, 148–152
  union-compatible, 123, 133
  unions of, 121
  viewing data, 49–55
  views. *See* views

text editors
  creating script files with, 202
  Oracle SQL *Plus, 45

titles, adding to reports, 209–210
totals, including in reports, 212–214
transactions, described, 145–146
TRTIM function, 200
TTITLE command, 209–210, 217
tuples in databases, 29

· ·

underscore (_) wildcard character, 99
undoing commands, 144–145
unions, using, 121–123
UNION command, 122–123, 132
union-compatible tables, 123, 133

UPDATE command, 55–56, 65, 140–142, 150, 153, 252, 255

updating
  constraining with conditions, 183
  cursors, 247
  data generally, 140
  data with INSERT command, 143–144
  databases through a view, 166–167
  rows, 252–253
  views, 169, 185

UPPER function, 194–195

· ·

values
  changing in columns, 55–56
  changing in columns to nulls, 147–148, 153
  and prompt variables, 231–232
  rounding off, 195–196

variables
  in PL/SQL, 255
  prompt, using, 231–232, 236
  strSQL string, 250

viewing table data, 49–55

views
  described, using, 160–170, 184
  involving statistics, 170
  using to query data for report, 204–205

Visual Basic, 249

· ·

WHERE clause
  described, using, 75, 77–78, 98
  and HAVING clause, 95–97
  in inner joins, 128–129
  and joins, 106, 108, 118–121

wildcards
  in Access, 84
  described, 99
  in MySQL, 83